T0155849

Lecture Notes in Artificial Intelligence   12705

Subseries of Lecture Notes in Computer Science

Series Editors

Randy Goebel
University of Alberta, Edmonton, Canada

Yuzuru Tanaka
Hokkaido University, Sapporo, Japan

Wolfgang Wahlster
DFKI and Saarland University, Saarbrücken, Germany

Founding Editor

Jörg Siekmann
DFKI and Saarland University, Saarbrücken, Germany

More information about this subseries at http://www.springer.com/series/1244

Manish Gupta · Ganesh Ramakrishnan (Eds.)

Trends and Applications in Knowledge Discovery and Data Mining

PAKDD 2021 Workshops
WSPA, MLMEIN, SDPRA, DARAI, and AI4EPT
Delhi, India, May 11, 2021
Proceedings

 Springer

Editors
Manish Gupta (iD)
Microsoft
Hyderabad, India

Ganesh Ramakrishnan
Indian Institute of Technology Bombay
Mumbai, India

ISSN 0302-9743   ISSN 1611-3349 (electronic)
Lecture Notes in Artificial Intelligence
ISBN 978-3-030-75014-5   ISBN 978-3-030-75015-2 (eBook)
https://doi.org/10.1007/978-3-030-75015-2

LNCS Sublibrary: SL7 – Artificial Intelligence

This Springer imprint is published by the registered company Springer Nature Switzerland AG
The registered company address is: Gewerbestrasse 11, 6330 Cham, Switzerland

Preface

The Pacific-Asia Conference on Knowledge Discovery and Data Mining (PAKDD) is one of the longest established and leading international conferences in the areas of data mining and knowledge discovery. It provides an international forum for researchers and industry practitioners to share their new ideas, original research results, and practical development experiences from all KDD-related areas, including data mining, data warehousing, machine learning, artificial intelligence, databases, statistics, knowledge engineering, visualization, decision-making systems, and emerging applications. PAKDD 2021 was held online during May 11–14, 2021.

Along with the main conference, PAKDD workshops intend to provide an international forum for researchers to discuss and share research results. After reviewing the workshop proposals, we were able to accept five workshops that covered topics in smart and precise agriculture, measurement informatics, scope detection of the peer review articles, data assessment and readiness for artificial intelligence, and enterprise process transformation. The diversity of topics in these workshops contributed to the main themes of the conference. There were, in all, 39 submissions across the five workshops. After carefully reviewing each submission, these workshops were able to accept 17 of the submissions as full papers. The five workshops were as follows:

- Workshop on Smart and Precise Agriculture (WSPA 2021)
- PAKDD 2021 Workshop on Machine Learning for MEasurement INformatics (MLMEIN 2021)
- The First Workshop and Shared Task on Scope Detection of the Peer Review Articles (SDPRA 2021)
- The First International Workshop on Data Assessment and Readiness for AI (DARAI 2021)
- The First International Workshop on Artificial Intelligence for Enterprise Process Transformation (AI4EPT 2021)

March 2021

Manish Gupta
Ganesh Ramakrishnan

Organization

Workshop Co-chairs

Manish Gupta	Microsoft, India
Ganesh Ramakrishnan	IIT Bombay, India

Workshop on Smart and Precise Agriculture (WSPA 2021)

Organizers

Sahely Bhadra	IIT Palakkad, India
Satyajit Das	IIT Palakkad, India
Mrinal Das	IIT Palakkad, India
Deepak Jaiswal	University of Illinois at Urbana-Champaign, USA

Program Committee

Adway Mitra	IIT Kharagpur, India
Amrita Roy	IIT Palakkad, India
Deepak Jaiswal	IIT Palakkad, India
Sreenath Vijaykumar	IIT Palakkad, India
Lakshmi N. Theagarajan	IIT Palakkad, India
Athira P	IIT Palakkad, India
Satyajit Das	IIT Palakkad, India
Sahely Bhadra	IIT Palakkad, India
Mrinal Das	IIT Palakkad, India

PAKDD 2021 Workshop on Machine Learning for MEasurement INformatics (MLMEIN 2021)

Organizer

Takashi Washio	Osaka University, Japan

Program Committee

Saso Dzeroski	Jozef Stefan Institute, Slovenia
Satoshi Hara	Osaka University, Japan
Tamiki Komatsuzaki	Hokkaido University, Japan
Hiromichi Nagao	University of Tokyo, Japan
Kenji Nagata	National Institute for Materials Science, Japan
Phi Le Nguyen	Hanoi University of Science and Technology, Vietnam
Shunsuke Ono	Tokyo Institute of Technology, Japan

Kai Ming Ting	Nanjing University, China
Ljupco Todorovski	University of Ljubljana, Slovenia

The First Workshop and Shared Task on Scope Detection of the Peer Review Articles (SDPRA 2021)

Organizers

Saichethan Miriyala Reddy	IIT Bhagalpur, India
Naveen Saini	Université Toulouse III – Paul Sabatier, France

Program Committee

Rajesh Piryani	South Asian University, India
Amit Kumar	Université de Caen Basse-Normandie, France
Paheli Bhattacharya	IIT Kharagpur, India
Jyoti Prakash Singh	NIT Patna, India
Nikiforos Pittaras	NCSR Demokritos, Greece
Judith Jeyafreeda	University of Caen Normandy, France
Silvio Moreira	North Eastern University, USA
Guillaume Cabanac	Université Toulouse III - Paul Sabatier, France
Roshni Chakraborty	Aalborg University, Denmark
Prasun Tripathi	IIT (ISM) Dhanbhad, India
Marc Bertin	Université Claude Bernard Lyon 1, France
Chifumi Nishioka	Kyoto University, Japan
Sayantan Mitra	IIT Patna, India
Aditya Joshi	CSIRO, Australia
Rohit Salgotra	Thapar University, India
Rakesh Sanodiya	IIT Patna, India

The First International Workshop on Data Assessment and Readiness for AI (DARAI 2021)

Organizers

Bortik Bandyopadhyay	Apple, USA
Sambaran Bandyopadhyay	IBM Research - India, India
Srikanta Bedathur	IIT Delhi, India
Nitin Gupta	IBM Research - India, India
Sameep Mehta	IBM Research - India, India
Shashank Mujumdar	IBM Research - India, India
Srinivasan Parthasarathy	Ohio State University, USA
Hima Patel	IBM Research - India, India

Program Committee

Shanmukha Guttula	IBM Research - India, India
Aniya Aggarwal	IBM Research - India, India

Pranay Lohia	IBM Research - India, India
Vitobha Munigala	IBM Research - India, India
Ruhi Sharma Mittal	IBM Research - India, India
Lokesh N	IIT Bombay, India
Naveen Panwar	IBM Research - India, India
Kishalay Das	IISC, India
Vishal Saley	IISC, India
Arushi Prakash	Amazon, India
Paarth Gupta	SMVDU, India

The First International Workshop on Artificial Intelligence for Enterprise Process Transformation (AI4EPT 2021)

Organizers

Monika Gupta	IBM Research - India, India
Kushal Mukherjee	IBM Research - India, India
Rakesh R. Pimplikar	IBM Research - India, India
Prerna Agarwal	IBM Research - India, India
Sampath Dechu	IBM Research - India, India
Shrihari Vasudevan	Ericsson, India

Program Committee

Vinod Muthusamy	IBM, USA
Pankaj Dhoolia	IBM Research - India, India
Tejas Dhamecha	IBM Research - India, India
Shivali Agarwal	IBM Research - India, India
Hagen Voelzer	IBM Research – Zurich, Switzerland
Ramasuri Narayanam	IBM Research - India, India
Christopher Butler	IBM Research – Australia, Australia
Rahul Purandare	Indraprastha Institute of Information Technology Delhi, India

Contents

**The First International Workshop on Data Assessment
and Readiness for AI (DARAI 2021)**

**The First International Workshop on Artificial Intelligence
for Enterprise Process Transformation (AI4EPT 2021)**

Workshop on Smart and Precise Agriculture (WSPA 2021)

Identification of Harvesting Year of Barley Seeds Using Near-Infrared Hyperspectral Imaging Combined with Convolutional Neural Network

Tarandeep Singh[1,2], Neerja Mittal Garg[1,2(✉)], and S. R. S. Iyengar[3]

[1] Academy of Scientific and Innovative Research, Ghaziabad 201002, India
[2] Computational Instrumentation, CSIR-CSIO, Chandigarh 160030, India
neerjamittal@csio.res.in
[3] Department of Computer Science, IIT Ropar, Rupnagar 140001, Punjab, India

Abstract. To evaluate the quality and safety of the seeds, identification of the harvesting year is one of the main parameters as the quality of the seeds is deteriorated during storage due to seed aging. In this study, hyperspectral imaging in the near-infrared range of 900–1700 nm was used to non-destructively identify the harvesting time of the barley seeds. The seeds samples including three years from 2017 to 2019 were collected. An end-to-end convolutional neural network (CNN) model was developed using the mean spectra extracted from the ventral and dorsal sides of the seeds. CNN model outperformed other classification models (K-nearest neighbors and support vector machines with and without spectral preprocessing) with a test accuracy of 97.25%. This indicated that near-infrared hyperspectral imaging combined with CNN could be used to rapidly and non-destructively identify the harvesting year of the barley seeds.

Keywords: Barley · Convolutional neural network · Harvesting year · Near-infrared hyperspectral imaging

1 Introduction

Barley (Hordeum vulgare L.) is one of the most widely consumed grains for both human consumption and animal feed. It is the preferred grain for preparing the malt due to the attached husk and better enzymatic activity. The quality of the seeds is dependent upon the chemical composition (such as protein and starch) which is deteriorated year by year. Furthermore, germination capabilities, plant growth, and grain yield are also affected with respect to the storage time of the seeds [1]. As it is difficult to accurately determine the storage time of the seeds with visual observation, some traders sell the mixture of seeds stored for years and the recently harvested seeds, which could lead to consumer rejection, and consequently affect sales and prices. Therefore, identification of the harvesting

© Springer Nature Switzerland AG 2021
M. Gupta and G. Ramakrishnan (Eds.): PAKDD 2021 Workshops, LNAI 12705, pp. 3–8, 2021.
https://doi.org/10.1007/978-3-030-75015-2_1

year of barley seeds using rapid, accurate, and non-destructive techniques has importance in the modern seed industry.

Near-infrared (NIR) Hyperspectral imaging (HSI) is a non-destructive and rapid technique that combines the traits of spectroscopy and imaging in a single system to jointly acquire the spatial and spectral information of a sample. HSI system acquires images in a three-dimensional "hypercube" comprising two spatial and one spectral dimension. Each pixel in a hyperspectral image is a complete spectrum. NIR-HSI technique is based upon the interaction of radiation with the sample i.e. combination vibrations and molecular overtone of N-H, O-H, and C-H bonds [2].

Recently, NIR-HSI has emerged as a widely used technique to assess the quality of various cereal grains including the classification of seeds from different years [1,3,4]. Different classifiers have been applied to analyze the hyperspectral data, i.e., soft independent modeling of class analogy, least squares support vector machine, and partial least squares discrimination analysis. Deep learning (DL) is a state-of-the-art machine learning model that has been widely applied in the field of NIR-HSI. Convolutional neural network (CNN) is one of the most successful DL models which extracts the abstract features by processing the input data through many layers [5]. This study aimed at the investigation of the feasibility of NIR-HSI combined with CNN to identify harvesting years of the barley seeds.

2   Materials and Methodology

2.1   Barely Samples and Near-Infrared Hyperspectral Imaging System

Barley seeds from three varieties (BH959, BHS352, and RD2552) and three harvesting years (2017, 2018, and 2019) were obtained from NBPGR, New Delhi. A total of 4536 (504 seeds from each variety and year) seeds were included in the study. In the present study, a push-broom (also called line scan) hyperspectral imaging system (Fig. 1) was used. The system incorporates a hyperspectral camera, spectrograph, lens, light source, translation stage, dark chamber to prevent stray light, and computer. The near-infrared hyperspectral camera used in the system collects the images in 168 wavelength bands ranging between 900–1700 nm with a spectral resolution of 4.9 nm. The samples of barley seeds were placed into 12 × 6 arrays of wells on an aluminum tray. Hypercubes were acquired for both the ventral (crease-up) and dorsal (crease-down) side of the kernels.

Image acquisition was performed using SpectrononPro software that controlled the speed and travel of the linear stage along with various imaging parameters e.g. integration time, frame rate, number of lines, and image correction. The resolution of the obtained hypercube is $320 \times 850 \times 164$ (x × y × λ), where x and y values refer to the spatial coordinates of the image whereas λ corresponds to the number of wavelengths. For image correction, the dark current response of the camera and white reference was obtained at an interval of one hour. The dark current response of the camera was acquired by placing a cap over the end

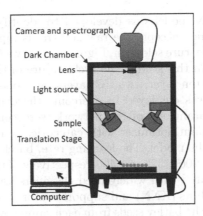

Fig. 1. Schematic diagram of the hyperspectral imaging system.

of the lens and white reference was acquired from a fluorilon tile with reflectivity >98%.

2.2 Model Development and Validation

Image processing and spectral data extraction were carried out (see Fig. 2(a) to (d)) before developing the classification models. A 3×3 median filter was applied to remove dead or bad pixels in the hypercube. From the obtained hyperspectral images, it was observed that an image at 1102.93 nm wavelength provided a good distinction between the barley kernels and the background. Therefore, the barley seeds were segmented from the background by taking a binary threshold for reflectance at 1102.93 nm >0.27. Subsequently, the mean spectrum was extracted from each single seed region of interest. Furthermore, due to poor signal-to-noise ratio, the wavelengths below 955.62 nm and beyond 1688.87 nm were truncated resulting in the hypercube between wavelength range 955.62–1688.87 nm (total 147 wavelengths).

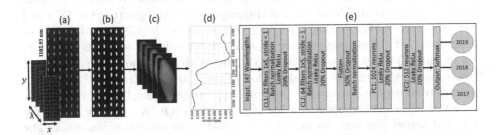

Fig. 2. Image segmentation, spectra extraction, and end-to-end CNN architecture: (a) hypercube showing image at 1102.93 nm; (b) image after threshold segmentation; (c) single seed ROI; (d) mean spectrum; (e) end-to-end CNN architecture.

An end-to-end CNN model was developed to identify the harvesting year of the barley seeds, which accepts the raw spectra as input without any spectral preprocessing and feature selection. Figure 2(e) shows CNN architecture and hyperparameters used in this study. The architecture of the CNN has two convolutional layers, a flatten layer, two FC layers, and one output layer. In the architecture, dropout was employed to overcome the overfitting problem (see Fig. 2(e)). The hyperparameters of the network were tuned to achieve higher validation accuracy. Adam optimizer with categorical cross-entropy loss function was used to train the model. The learning rate, batch size, and the number of epochs were set to 0.0001, 32, and 400 respectively. The performance of the CNN model was compared with widely used classifiers for spectral data analysis i.e., K-nearest neighbors (KNN) and support vector machines (SVM). Before developing the models, the barley seeds from each sample were randomly divided into the training and the testing set with a ratio of 8:2.

SVM [6] and KNN [7] are widely used supervised machine learning algorithms for spectral data analysis. In SVM, the multidimensional hyperplanes are used to maximally separate the classes whereas in KNN unknown samples are assigned a class by selecting the most frequent class belonging to their 'K' number of closest samples. In the case of SVM, RBF was used as the kernel function. Before feeding the spectra to SVM and KNN models, different spectral pre-treatments were applied. In this study, standard normal variate (SNV), multiplicative scatter correction (MSC), Savitzky-Golay (SG) smoothing, SG first derivative, SG second derivative, and detrending were carried out [2].

In this study, Spyder 4.1.3 IDE was used to write the custom code in Python 3.7 programming language. The open-source libraries namely 'Spectral', 'Scikit-image', 'OpenCV', 'Numpy', Scikit-learn, Scipy, and Keras were used. The experiments were carried out on an HP Z8 workstation with 64 GB of RAM and an 8 GB Nvidia Quadro P4000 graphics card.

3 Results and Discussion

To build the different classification models, mean spectra were extracted from the ventral side and dorsal side of the seeds (total 4536 seeds) resulting in a total of 9072 mean spectra. The number of training and testing spectra were 7254 and 1818 respectively. For tuning the hyperparameters of the CNN model, 20% of the training set (1458 spectra) were chosen as the validation set. This procedure was repeated 5 times to check the repeatability with the same hyperparameters. Subsequently, mean ± standard deviation of the 5 random train and validation splits were considered to select the optimal hyperparameters. In the case of KNN and SVM, to select the optimal value of parameters (i.e. 'K', penalty parameter (C), and kernel function parameter (γ)), 5-fold cross-validation was used. The training and validation accuracy of the models in case of the optimal parameters is shown in Table 1.

After selecting the optimal parameters, all the models were trained and tested with the same training (7254 spectra) and testing set (1818 spectra) to make a

Table 1. Training and validation accuracy for optimal model parameters.

Model	Training accuracy (%)	Validation accuracy (%)
CNN	99.94 ± 0.06	96.75 ± 0.25
SVM	100 ± 0.0	93.37 ± 0.64
KNN	86.43 ± 0.31	80.26 ± 0.53

fair comparison. The end-to-end CNN model accepts raw spectra without spectral preprocessing as well as feature selection, whereas raw and preprocessed spectra were fed into SVM and KNN model. The performance of the three models is shown in Table 2. The CNN model outperformed other classifiers and achieved an accuracy of 97.25%. Likewise, the number of correctly classified spectra, precision, recall, and F1 score were greater as compared to other classifiers (see Fig. 3). The deep spectral features extracted by convolutional layers were easily classified as compared to the original spectra which led to better performance of the CNN model.

Table 2. Performance of different classification models.

Model	Parameters[a]	Training accuracy (%)	Testing accuracy (%)
CNN	(2, 5, 32, 400)	99.92	97.25
SVM[b]	(100, 0.01)	100	92.52
KNN[b]	7	87.17	80.91

[a]Parameters of different discriminant models: number of convolutional layers, kernel size, batch size, and epoch for CNN model; number of neighbors for KNN; penalty parameter (C) and kernel function parameter (γ) for SVM model

[b]Classifier combined with Savitzky-Golay second derivative

Actual class	Confusion matrix			Precision	Recall	F1 score
2019	589	11	6	99.33	97.19	98.25
2019	575	12	19	92.89	94.88	93.88
2019	552	19	35	82.39	91.09	86.52
2018	2	592	12	95.48	97.69	96.57
2018	13	566	27	92.48	93.4	92.94
2018	36	518	52	78.48	85.48	81.83
2017	2	17	587	97.02	96.86	96.94
2017	31	34	541	92.16	89.27	90.7
2017	82	123	401	82.17	66.17	73.31
	2019	2018	2017			

Predicted class   Color coding: CNN, SVM, KNN

Fig. 3. Confusion matrix and classification report of different models.

4   Conclusions

In this study, a non-destructive approach for the identification of the harvesting year of barley seeds using near-infrared hyperspectral imaging (900–1700 nm) has been demonstrated. An end-to-end CNN model was developed to classify the spectra extracted from the ventral and the dorsal side of the seeds. KNN and SVM models with six different spectral pre-processing techniques were compared with the CNN model. CNN outperformed other models and achieved an accuracy of 97.25%. The results indicated that near-infrared hyperspectral imaging coupled with a convolutional neural network has great potential to identify the harvesting year of barley seeds. In the future, barley seeds from different locations and more years can be collected to build a robust model against different climatic and environmental conditions.

References

1. Gao, D., Jian, G., Cheng, W., et al.: Differentiation of storage time of wheat seed based on near infrared hyperspectral imaging. Int. J. Agric. Biol. Eng. **10**, 251–258 (2017). https://doi.org/10.3965/j.ijabe.20171002.1619
2. Jia, B., Wang, W., Ni, X., et al.: Essential processing methods of hyperspectral images of agricultural and food products. Chemom. Intell. Lab. Syst. **198** (2020). https://doi.org/10.1016/j.chemolab.2020.103936
3. Wang, Q., Huang, M., Zhu, Q.: Characteristics of maize endosperm and germ in the geographical origins and years identification using hyperspectral imaging. In: 2014 ASABE Annual International Meeting. American Society of Agricultural and Biological Engineers, pp. 1–6 (2014)
4. Guo, D., Zhu, Q., Huang, M., et al.: Model updating for the classification of different varieties of maize seeds from different years by hyperspectral imaging coupled with a pre-labeling method. Comput. Electron. Agric. **142**, 1–8 (2017). https://doi.org/10.1016/j.compag.2017.08.015
5. Yang, J., Xu, J., Zhang, X., et al.: Deep learning for vibrational spectral analysis: recent progress and a practical guide. Anal. Chim. Acta **1081**, 6–17 (2019). https://doi.org/10.1016/j.aca.2019.06.012
6. Hearst, M.A., Dumais, S.T., Osuna, E., et al.: Support vector machines. IEEE Intell. Syst. Appl. **13**, 18–28 (1998)
7. Peterson, L.E.: K-nearest neighbor. Scholarpedia **4**, 1883 (2009)

Plant Leaf Disease Segmentation Using Compressed UNet Architecture

Mohit Agarwal[✉], Suneet Kr. Gupta, and K. K. Biswas

Department of CSE, Bennett University, Greater Noida, India
ma8573@bennett.edu.in

Abstract. In proposed work, a compressed version of UNet has been developed using Differential Evolution for segmenting the diseased regions in leaf images. The compressed model has been evaluated on potato late blight leaf images from PlantVillage dataset. The compressed model needs only 6.8% of space needed by original UNet architecture, and the inference time for disease classification is twice as fast without loss in performance metric of mean Intersection over Union (IoU).

Keywords: UNet architecture · Plant disease classification · Differential evolution · Optimization · Compression and acceleration

1 Introduction and Literature Survey

Several recent research works have been proposed for identifying plant diseases from leaf images [10, 15, 16]. Moreover, researchers have also tried to segment diseased region using Machine Learning [9, 11, 22] and Deep Learning methods [5, 13] for identification of extent of the disease. In 2015 Ronneberger et al. [17] have proposed UNet, a highly efficient encoder-decoder Convolution Neural Network (CNN) architecture for pixel wise segmentation of the images. However, it is very difficult to deploy the UNet on edge computing devices i.e. smart camera, raspberry-pi etc. as these devices have a limited memory and computational power. There is thus a need for developing compressed versions of such models so that these can be utilized in edge computing devices.

In the last decade, the research community has applied different techniques such as matrix factorization, quantization, flattened convolution, network pruning, huffman coding etc. for compression of CNN models [1, 6, 7, 12, 14, 21]. More recently meta-heuristic approaches have also been also used for compression of CNN models [18–20]. Beheshti and Johnsson [2] have squeezed the UNet architecture and experimentally demonstrated that the squeezed version of UNet required 12.08x times less memory and 1.48x times faster inference time without compromising the performance evaluation metric.

In this paper, a novel differential evolution [4] based approach has been proposed for minimizing the number of filters in convolution layers of UNet architecture without compromising the performance evaluation metric of mean Intersection over Union. Moreover, the compressed UNet model can be utilized in farm

M. Gupta and G. Ramakrishnan (Eds.): PAKDD 2021 Workshops, LNAI 12705, pp. 9–14, 2021.
https://doi.org/10.1007/978-3-030-75015-2_2

rover robots for segmenting the diseased region of leaves in captured images, and to spray fungicide on affected areas.

The main contribution of this paper is design of a novel fitness function for compressing UNet architecture using Differential Evolution for segmenting diseased region in plant leaves.

The rest of the paper is organized as follows: The proposed model is described in Sect. 2. The experimental setup and results are discussed in Sect. 3, followed by conclusion in Sect. 4.

2   Proposed Model

UNet is an encoder-decoder type deep learning network (Fig. 4 in Appendix A.1) which consist of 23 convolution layers. The encoder part has 10 convolution layers in two sets with 64, 128, 256, 512 and 1024 kernels. A max-pooling layer follows each set of convolution layer. The decoder part has 13 convolution layers, the last layer being a softmax layer.

We propose to compress the UNet model using Differential Evolution (DE) [3, 4] by retaining the most relevant filters and nodes at each layer. The flow chart of Differential Evolution has been depicted in Fig. 1.

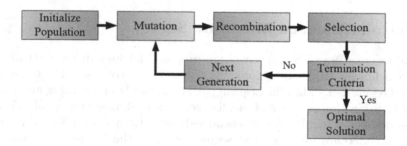

Fig. 1. Differential Evolution flow chart.

In differential evolution, there are mainly three intermediate steps 1) Mutation, 2) Recombination, and 3) Selection which are performed on initial population. In our case the population was created by a random binary vector representing the nodes in various layers of UNet model (sample vector is shown in Fig. 2). The length of this vector is 6848 representing the total number of convolution filters in UNet.

In the first operation of mutation, three vectors are randomly chosen from initial population and a new vector is generated after adding one of the vectors to the difference of remaining two vectors multiplied by a mutation factor lying in the range 0 to 1. The vector so generated (donor vector) after mutation operation was passed for recombination operation where it shares the information with target vector in that pool. In recombination operation, a random number is generated between 0 and 1 for each index position of vectors, and if it is found

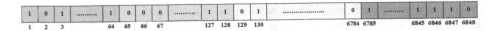

Fig. 2. A sample vector randomly initialized with 0's and 1's representing various filters in UNet.

to be greater than a chosen recombination factor (taken as 0.7), then the trial vector element at that index position is taken from the target vector, else it is taken from donor vector. The goodness of the generated trial vector is dependent on the fitness value which is computed in the selection step of DE explained in next section.

2.1   Selection in DE

The selection process is a compromise between α_j the fraction of nodes retained at layer j, and the accuracy metric β at the kth compression stage. β may be chosen as mean Intersection over Union performance metric. The objective function chosen is:

$$max(w \times (1 - \alpha_j) + (1 - w) \times \beta) \ subject \ to \ \alpha_j \leq 1 \qquad (1)$$

where w is a relative weighting factor between the two sub-objectives. A higher choice of w would lead to a large reduction in filters at each layer, but at the cost of compromising accuracy, while a smaller choice of w would ensure higher accuracy at the cost of lower model compression.

For significant model compression, w may be chosen as 0.5 or more, and all steps of DE are repeated till the accuracy metric deteriorates by a preset acceptable threshold.

3   Experimental Setup and Results

The experiments were performed using NVIDIA DGX v100 machine on UBUNTU 20.0.1 operating system with python programming language. The compressed UNet was used for disease identification in potato crop late blight leaf images from PlantVillage dataset [8]. To create ground truth images (refer 2nd row in Fig. 3) the diseased regions were manually annotated and marked with brown color by using RGB value (128, 128, 0). Moreover, a part of the leaf image similar to green color was marked green with RGB value (0, 255, 0). The background was assigned black color with RGB value (0, 0, 0). The data was transformed in HSV space, and Hue value was checked for green, brown and yellow parts of the image.

The original UNet model needed 121,303 KB storage space. Since UNet weights were randomly initialized and DE initial vectors were also random, the model alongwith compression was executed 5 times and mIoU was recorded before and after compression. Model size after compression was also recorded as shown in Table 1. From the mean of 5 executions it was found that 15.49x times space reduction could be achieved with mIoU gain of around 1.62%. The gain in mIoU can be attributed to better choice of filters forced by compression of nodes. The visual comparison of ground truth and predicted outputs for single execution are depicted in Fig. 3.

Moreover, after compression of the UNet model the number of floating point operation reduces considerably, which enhances the inference time on test data by 2.30x times.

Table 1. Comparison of performance and size for different executions of compressing UNet.

Experiment No.	mIoU (Before Compression)	mIoU (After Compression)	Compressed size
1	92.34%	94.29%	7,668 KB
2	92.56%	94.88%	8,223 KB
3	93.33%	94.67%	8,129 KB
4	93.49%	93.56%	7,492 KB
5	91.48%	93.92%	7,623 KB
Mean	92.64%	94.26%	7,827 KB

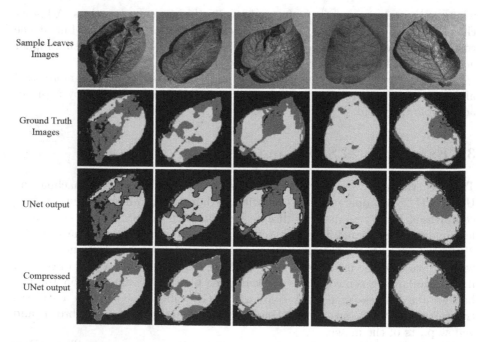

Fig. 3. Sample images of dataset, ground truth annotations of same images, predicted outputs of UNet, predicted output of compressed UNet.

4   Conclusion

This paper has proposed a scheme for developing a compressed UNet architecture using differential evolution, so that it can be deployed on edge devices. It has been shown that without compromising on mean IoU, the storage space is reduced by 15.49x times and inference time for predicting the diseased region is improved by 2.30x times.

A   Appendix I

A.1   UNet architecture

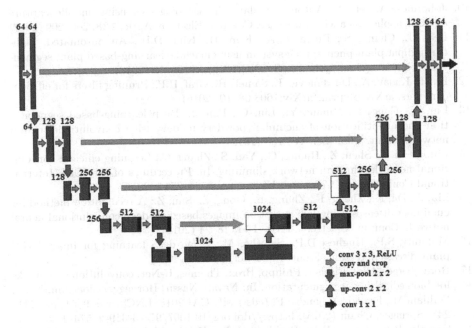

Fig. 4. Architecture of UNet [17].

References

1. Anwar, S., Hwang, K., Sung, W.: Structured pruning of deep convolutional neural networks. ACM J. Emerg. Technol. Comput. Syst. (JETC) **13**(3), 1–18 (2017)
2. Beheshti, N., Johnsson, L.: Squeeze u-net: a memory and energy efficient image segmentation network. In: Proceedings of the IEEE/CVF Conference on Computer Vision and Pattern Recognition Workshops, pp. 364–365 (2020)
3. Chakraborty, U.K.: Advances in Differential Evolution, vol. 143. Springer, Heidelberg (2008). https://doi.org/10.1007/978-3-540-68830-3
4. Feoktistov, V.: Differential Evolution. Springer, Heidelberg (2006). https://doi.org/10.1007/978-0-387-36896-2
5. Ganesh, P., Volle, K., Burks, T., Mehta, S.: Deep orange: mask R-CNN based orange detection and segmentation. IFAC-PapersOnLine **52**(30), 70–75 (2019)

6. Han, S., Pool, J., Tran, J., Dally, W.: Learning both weights and connections for efficient neural network. In: Advances in Neural Information Processing Systems, pp. 1135–1143 (2015)
7. He, Y., Zhang, X., Sun, J.: Channel pruning for accelerating very deep neural networks. In: Proceedings of the IEEE International Conference on Computer Vision, pp. 1389–1397 (2017)
8. Hughes, D., Salathé, M., et al.: An open access repository of images on plant health to enable the development of mobile disease diagnostics. arXiv preprint arXiv:1511.08060 (2015)
9. Islam, M., Dinh, A., Wahid, K., Bhowmik, P.: Detection of potato diseases using image segmentation and multiclass support vector machine. In: 2017 IEEE 30th Canadian Conference on Electrical and Computer Engineering (CCECE), pp. 1–4. IEEE (2017)
10. Johannes, A., et al.: Automatic plant disease diagnosis using mobile capture devices, applied on a wheat use case. Comput. Electron. Agric. **138**, 200–209 (2017)
11. Lee, U., Chang, S., Putra, G.A., Kim, H., Kim, D.H.: An automated, high-throughput plant phenotyping system using machine learning-based plant segmentation and image analysis. PLoS ONE **13**(4), e0196615 (2018)
12. Li, H., Kadav, A., Durdanovic, I., Samet, H., Graf, H.P.: Pruning filters for efficient convnets. arXiv preprint arXiv:1608.08710 (2016)
13. Lin, K., Gong, L., Huang, Y., Liu, C., Pan, J.: Deep learning-based segmentation and quantification of cucumber powdery mildew using convolutional neural network. Front. Plant Sci. **10**, 155 (2019)
14. Liu, Z., Li, J., Shen, Z., Huang, G., Yan, S., Zhang, C.: Learning efficient convolutional networks through network slimming. In: Proceedings of the IEEE International Conference on Computer Vision, pp. 2736–2744 (2017)
15. Ma, J., Du, K., Zheng, F., Zhang, L., Gong, Z., Sun, Z.: A recognition method for cucumber diseases using leaf symptom images based on deep convolutional neural network. Comput. Electron. Agric. **154**, 18–24 (2018)
16. Mohanty, S.P., Hughes, D.P., Salathé, M.: Using deep learning for image-based plant disease detection. Front. Plant Sci. **7**, 1419 (2016)
17. Ronneberger, Olaf, Fischer, Philipp, Brox, Thomas: U-Net: convolutional networks for biomedical image segmentation. In: Navab, Nassir, Hornegger, Joachim, Wells, William M., Frangi, Alejandro F. (eds.) MICCAI 2015. LNCS, vol. 9351, pp. 234–241. Springer, Cham (2015). https://doi.org/10.1007/978-3-319-24574-4_28
18. Samala, R.K., Chan, H.P., Hadjiiski, L.M., Helvie, M.A., Richter, C., Cha, K.: Evolutionary pruning of transfer learned deep convolutional neural network for breast cancer diagnosis in digital breast tomosynthesis. Phys. Med. Biol. **63**(9), 095005 (2018)
19. Wang, Z., Li, F., Shi, G., Xie, X., Wang, F.: Network pruning using sparse learning and genetic algorithm. Neurocomputing **404**, 247–256 (2020)
20. Yang, Chuanguang, An, Zhulin, Li, Chao, Diao, Boyu, Xu, Yongjun: Multi-objective pruning for CNNs using genetic algorithm. In: Tetko, Igor V., Kůrková, Věra, Karpov, Pavel, Theis, Fabian (eds.) ICANN 2019. LNCS, vol. 11728, pp. 299–305. Springer, Cham (2019). https://doi.org/10.1007/978-3-030-30484-3_25
21. Zhang, Q., Zhang, M., Chen, T., Sun, Z., Ma, Y., Yu, B.: Recent advances in convolutional neural network acceleration. Neurocomputing **323**, 37–51 (2019)
22. Zhou, J., Fu, X., Zhou, S., Zhou, J., Ye, H., Nguyen, H.T.: Automated segmentation of soybean plants from 3D point cloud using machine learning. Comput. Electron. Agric. **162**, 143–153 (2019)

PAKDD 2021 Workshop on Machine Learning for MEasurement INformatics (MLMEIN 2021)

Hierarchical Topic Model for Tensor Data and Extraction of Weekly and Daily Patterns from Activity Monitor Records

Shunichi Nomura[1,2]([⊠]), Michiko Watanabe[3], and Yuko Oguma[3]

[1] The Institute of Statistical Mathematics, Tokyo, Japan
[2] Waseda University, Tokyo, Japan
snomura5@aoni.waseda.jp
[3] Keio University, Kanagawa, Japan

Abstract. Latent Dirichlet allocation (LDA) is a popular topic model for extracting common patterns from discrete datasets. It is extended to the pachinko allocation model (PAM) with a hierarchical topic structure. This paper presents a combination meal allocation (CMA) model, which is a further enhanced topic model from the PAM that has both hierarchical categories and hierarchical topics. We consider count datasets in multiway arrays, i.e., tensors, and introduce a set of topics to each mode of the tensors. The topics in each mode are interpreted as patterns in the topics and categories in the next mode. Despite there being a vast number of combinations in multilevel categories, our model provides simple and interpretable patterns by sharing the topics in each mode. Latent topics and their membership are estimated using Markov chain Monte Carlo (MCMC) methods. We apply the proposed model to step-count data recorded by activity monitors to extract some common activity patterns exhibited by the users. Our model identifies four daily patterns of ambulatory activities (commuting, daytime, nighttime, and early-bird activities) as sub-topics, and six weekly activity patterns as super-topics. We also investigate how the amount of activity in each pattern dynamically affects body weight changes.

Keywords: Topic model · Tensor data · Activity monitor records

1 Introduction

Topic models are clustering methods that are mainly applied to textual data for document classification, text mining, and other related tasks. In latent Dirichlet allocation (LDA) [3], documents are first transformed into "bags of words" expressions, i.e., frequency tables of the words. Then, topic models identify several multinomial distributions over the words shared across the documents as "topics." After the inference procedure, each document is represented by a mixture of "topics," and the words in it are regarded as a sample from the mixture of multinomial distributions corresponding to the identified topics. Topic models

© Springer Nature Switzerland AG 2021
M. Gupta and G. Ramakrishnan (Eds.): PAKDD 2021 Workshops, LNAI 12705, pp. 17–30, 2021.
https://doi.org/10.1007/978-3-030-75015-2_3

are successful models in textual analysis and also applied to non-textual data such as diagnosis and medication [10], purchase [5], and traffic speed data [12]. LDA is a fundamental model in topic analysis and some extended models with side information [2,15], temporal correlation [6], and hierarchical structures [9] have been proposed.

Herein, we propose a combination meal allocation (CMA) model, a hierarchical topic model for count-valued tensor data. We assume that the modes of the tensor are in a hierarchical order such that for each count, the category in a mode affects those in the subsequent modes. For instance, let us consider the ordering data of a combination meal in a fast-food restaurant. The combination meal consists of the main dish, side dish, and drink chosen from the menu. Every order is counted in an element of the tensor data, which have three modes for the choice of the main dish, side dish, and drink. Typically, we choose the main dish, side dish, and drink in turn, and each choice depends on the preceding choices and preference of the orderer. The CMA model is a multilevel topic model that considers such a type of hierarchy by associating hierarchical topic layers with the modes of the tensor data.

The generative model of the CMA model is illustrated in Fig. 1. The topics in each layer represent the multinomial distribution of the topics and categories one level down and, therefore, are interpreted as patterns in the lower levels such as side dishes and drinks. Compared to ordinary hierarchical models, such as a pachinko allocation model (PAM) [9], the CMA model reduces complexity and provides interpretable results by sampling multilevel categories in turn and sharing the same topics for each level. Hyperparameters in Dirichlet prior distributions and latent topics for respective counts are estimated using the collapsed Gibbs sampler [11].

We apply the proposed model to hourly step-count data recorded by activity monitors. These data are split into one-week sets, from Monday to Sunday, and arranged in arrays based on the hours of the day and days of the week. Our model consists of two topic layers with super-topics as the weekly patterns and sub-topics as the daily patterns of ambulatory activities. All step counts are allocated to the extracted activity patterns through the MCMC method. Using the results, we discuss the differences in weight reduction effects among weekly activity patterns.

2 Model

Consider L-th order tensor data $M_1, \ldots, M_D$ of size $W_1 \times \cdots \times W_L$ whose elements are nonnegative integers, i.e., count data. We denote the categories in the respective modes by $\mathcal{W}_1 = \{1, \ldots, W_1\}, \ldots, \mathcal{W}_L = \{1, \ldots, W_L\}$. Let $N_1, \ldots, N_D$ be the total counts of tensors $M_1, \ldots, M_D$. For convenience, we transform each tensor M_d into a sequence of vectors $v_{d1} = (v_{d1}^{(1)}, \ldots, v_{d1}^{(L)}), \ldots, v_{dN_d} = (v_{dN_d}^{(1)}, \ldots, v_{dN_d}^{(L)}) \in \mathcal{W}_1 \times \cdots \times \mathcal{W}_L$, as mode-wise categories of individual counts. We introduce multilevel topic layers $\mathcal{K}_0 = \{1, \ldots, K_0\}, \ldots, \mathcal{K}_{L-1} = \{1, \ldots, K_{L-1}\}$, over categories $\mathcal{W}_1, \ldots, \mathcal{W}_L$ of respective modes. Latent topics are allocated to the respective counts and denoted by $z_{d1} = (z_{d1}^{(0)}, \ldots, z_{d1}^{(L-1)}), \ldots, z_{dN_d} = (z_{dN_d}^{(0)}, \ldots, z_{dN_d}^{(L-1)}) \in \mathcal{K}_0 \times \cdots \times \mathcal{K}_{L-1}$, for each tensor data M_d. At each layer $l = 1, \ldots, L - 1$, its parent topics $1, \ldots, K_{l-1} \in \mathcal{K}_{l-1}$ represent multinomial distributions over the product set $\mathcal{K}_l \times \mathcal{W}_l$ of the topics and categories. For each dataset $d = 1, \ldots, D$, we define $\theta_d^{(0)} \in \Theta^{(0)} = \{\theta \in [0,1]^{K_0}; \sum_{k=1}^{K_0} \theta_k = 1\}$ and $\theta_{d1}^{(l)}, \ldots, \theta_{dK_{l-1}}^{(l)} \in \Theta^{(l)} = \{\theta \in [0,1]^{K_l \times W_l}; \sum_{k=1}^{K_l} \sum_{w=1}^{W_l} \theta_{kw} = 1\}$ for $l = 1, \ldots, L - 1$ as the probability mass of multinomial distributions for the topics. For the bottom layer, its parent topics $1, \ldots, K_{L-1} \in \mathcal{K}_{L-1}$ represent multinomial distributions over the last categories $\mathcal{W}_L$ and have a common probability mass $\theta_1^{(L)}, \ldots, \theta_{K_{L-1}}^{(L)} \in \Theta^{(L)} = \{\theta \in [0,1]^{W_L}; \sum_{w=1}^{W_L} \theta_w = 1\}$ among all the data. Furthermore, we introduce the Dirichlet distribution as a prior distribution for the probability mass of multinomial distributions.

The CMA model assumes that count datasets are generated by the following two-step process:

1. Sample sets of probability mass $\theta^{(0)}, \ldots, \theta^{(L)}$ over topics and categories in each layer as follows:
   - In the top layer, sample $\theta_d^{(0)} \sim \text{Dirichlet}(\alpha^{(0)})$ for each dataset $d = 1, \ldots, D$.
   - In the middle layers $l = 1, \ldots, L-1$, sample $\theta_{dk}^{(l)} \sim \text{Dirichlet}(\alpha_k^{(l)})$ for each dataset $d = 1, \ldots, D$ and parent-topic $k = 1, \ldots, K_{l-1}$.
   - In the bottom layer, sample $\theta_k^{(L)} \sim \text{Dirichlet}(\alpha^{(L)})$ for each parent-topic $k = 1, \ldots, K_{L-1}$.
2. For each count $n = 1, \ldots, N_d$ of each dataset $d = 1, \ldots, D$, sample its topics $z_{dn}^{(0)}, \ldots, z_{dn}^{(L-1)}$ and categories $v_{dn}^{(1)}, \ldots, v_{dn}^{(L)}$ in each layer as follows:
   - In the top layer, sample $z_{dn}^{(0)} \sim \text{Discrete}(\theta_d^{(0)})$.
   - In the middle layers $l = 1, \ldots, L-1$, sample $(z_{dn}^{(l)}, v_{dn}^{(l)}) \sim \text{Discrete}(\theta_{dz_{dn}^{(l-1)}}^{(l)})$ in order.
   - In the bottom layer, sample $v_{dn}^{(L)} \sim \text{Discrete}(\theta_{z_n^{(L-1)}}^{(L)})$.

In this process, the sets of multinomial parameters $\theta^{(0)} \in \mathbb{R}^D \times \Theta^{(0)}, \theta^{(1)} \in \mathbb{R}^{D \times K_0} \times \Theta^{(1)}, \ldots, \theta^{(L-1)} \in \mathbb{R}^{D \times K_{L-2}} \times \Theta^{(L-1)}, \theta^{(L)} \in \mathbb{R}^{K_{L-1}} \times \Theta^{(L)}$ are sampled from Dirichlet distributions with the sets of hyperparameters $\alpha^{(0)} \in \mathbb{R}^{K_0}, \alpha^{(1)} \in \mathbb{R}^{K_0 \times K_1 \times W_1}, \ldots, \alpha^{(L-1)} \in \mathbb{R}^{K_{L-2} \times K_{L-1} \times W_{L-1}}, \alpha^{(L)} \in \mathbb{R}^{W_L}$, respectively. This generative process can be represented by the graphical model in Fig. 2. If there is only one category for each of mode-1 to mode-$(L - 1)$, the CMA model is reduced to PAM [9]. By omitting the topics and categories in all the middle layers, the CMA model is reduced to LDA [3].

The conditional probabilities of topics $z_d = \{z_{d1}, \ldots, z_{dN_d}\}$ and categories $v_d = \{v_{d1}, \ldots, v_{dN_d}\}$ for each dataset $d = 1, \ldots, D$, given the multinomial parameters $\theta_d^{(0)}, \ldots, \theta_d^{(L-1)}, \theta^{(L)}$, are decomposed into

$$p(z_d, v_d|\theta_d) = p(z_d^{(0)}|\theta_d^{(0)}) \left\{ \prod_{l=1}^{L-1} p(z_d^{(l)}, v_d^{(l)}|z_d^{(l-1)}, \theta_d^{(l)}) \right\} p(v_d^{(L)}|z_d^{(L-1)}, \theta^{(L)}) \quad (1)$$

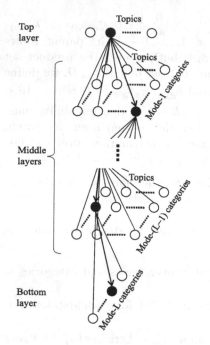

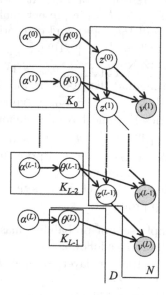

Fig. 1. Generative process of CMA model

Fig. 2. Graphical model of CMA model

where $z_d^{(l)} = \{z_{d1}^{(l)}, \ldots, z_{dN_d}^{(l)}\}$ for $l = 0, \ldots, L-1$, and $v_d^{(l)} = \{v_{d1}^{(l)}, \ldots, v_{dN_d}^{(l)}\}$ for $l = 1, \ldots, L$. Then, for each layer, the multinomial probabilities can be integrated out, and marginal probabilities for topics $z^{(l)} = \{z_1^{(l)}, \ldots, z_D^{(l)}\}$ and categories $v^{(l)} = \{v_1^{(l)}, \ldots, v_D^{(l)}\}$ in all the datasets can be obtained by

$$p(z^{(0)}|\alpha^{(0)}) = \prod_{d=1}^{D} \int_{\Theta^{(0)}} p(z_d^{(0)}|\theta_d^{(0)})\pi(\theta_d^{(0)}|\alpha^{(0)})\mathrm{d}\theta_d^{(0)}$$

$$= \prod_{d=1}^{D} \frac{\Gamma(\alpha_\bullet^{(0)})}{\Gamma(N_d + \alpha_\bullet^{(0)})} \prod_{k=1}^{K_0} \frac{\Gamma(N_{dk}^{(0)} + \alpha_k^{(0)})}{\Gamma(\alpha_k^{(0)})} \qquad (2)$$

for the top layer,

$$p(z^{(l)}, v^{(l)}|z^{(l-1)}, \alpha^{(l)}) = \prod_{d=1}^{D} \int_{\Theta^{(l)}} p(z_d^{(l)}, v_d^{(l)}|z_d^{(l-1)}, \theta_d^{(l)})\pi(\theta_d^{(l)}|\alpha^{(l)})\mathrm{d}\theta_d^{(l)}$$

$$= \prod_{d=1}^{D} \prod_{k=1}^{K_{l-1}} \frac{\Gamma(\alpha_{k\bullet\bullet}^{(l)})}{\Gamma(N_{dk}^{(l-1)} + \alpha_{k\bullet\bullet}^{(l)})} \prod_{k'=1}^{K_l} \prod_{w=1}^{W_l} \frac{\Gamma(N_{dkk'w}^{(l-1:l)} + \alpha_{kk'w}^{(l)})}{\Gamma(\alpha_{kk'w}^{(l)})}$$

$$\qquad (3)$$

for the middle layers $l = 1, \ldots, L - 1$, and

$$p(v^{(L)}|z^{(L-1)}, \alpha^{(L)}) = \int_{\Theta^{(L)}} \left\{ \prod_{d=1}^{D} p(v_d^{(L)}|z_d^{(L-1)}, \theta^{(L)}) \right\} \pi(\theta^{(L)}|\alpha^{(L)}) d\theta^{(L)}$$

$$= \prod_{k=1}^{K_{L-1}} \frac{\Gamma(\alpha_\bullet^{(L)})}{\Gamma(N_k^{(L-1)} + \alpha_\bullet^{(L)})} \prod_{w=1}^{W_L} \frac{\Gamma(N_{kw}^{(L-1:L)} + \alpha_w^{(L)})}{\Gamma(\alpha_w^{(L)})} \quad (4)$$

for the bottom layer, where Γ is a gamma function, $\alpha_\bullet^{(0)} = \sum_{k=1}^{K_0} \alpha_k^{(0)}$, $\alpha_{k\bullet\bullet}^{(l)} = \sum_{k'=1}^{K_l} \sum_{w=1}^{W_l} \alpha_{kk'w}^{(l)}$, $\alpha_\bullet^{(L)} = \sum_{w=1}^{W_L} \alpha_w^{(L)}$ and

$$N_{dk}^{(l-1)} = \sum_{n=1}^{N_d} I(z_{dn}^{(l)} = k), \quad (5)$$

$$N_{dkk'w}^{(l-1:l)} = \sum_{n=1}^{N_d} I(z_{dn}^{(l-1)} = k, z_{dn}^{(l)} = k', v_{dn}^{(l)} = w), \quad (6)$$

$$N_k^{(L-1)} = \sum_{d=1}^{D} \sum_{n=1}^{N_d} I(z_{dn}^{(L-1)} = k), \quad (7)$$

$$N_{kw}^{(L-1:L)} = \sum_{d=1}^{D} \sum_{n=1}^{N_d} I(z_{dn}^{(L-1)} = k, v_{dn}^{(L)} = w) \quad (8)$$

are the total counts of the corresponding topics and categories defined using an indicator function I.

3 Parameter Inference

For hierarchical topic models, an MCMC method called collapsed Gibbs sampling [11] can be used to simultaneously estimate latent topics and parameters. In sampling algorithms, the multinomial parameters $\theta^{(0)}, \ldots, \theta^{(L)}$ are omitted for computational efficiency by using the marginal probabilities in Eqs. (2), (3), and (4). The conditional probability for updating the topics of each count is given by

$$p(z_{dn} = (k_0, \ldots, k_{L-1})|z_{\backslash dn}, v, \alpha)$$
$$\propto p(z_{dn} = (k_0, \ldots, k_{L-1}), v_{dn}|z_{\backslash dn}, v_{\backslash dn}, \alpha)$$
$$\propto (N_{dk_0\backslash dn}^{(0)} + \alpha_{k_0}^{(0)}) \left\{ \prod_{l=1}^{L-1} \frac{N_{dk_{l-1}k_l v_{dn}^{(l)}\backslash dn}^{(l-1:l)} + \alpha_{k_{l-1}k_l v_{dn}^{(l)}}^{(l)}}{N_{dk_{l-1}\backslash dn}^{(l-1)} + \alpha_{k_{l-1}\bullet\bullet}^{(l)}} \right\} \frac{N_{k_{L-1}v_{dn}^{(L)}\backslash dn}^{(L-1:L)} + \alpha_{v_{dn}^{(L)}}^{(L)}}{N_{k_{L-1}\backslash dn}^{(L-1)} + \alpha_\bullet^{(L)}}$$
$$\quad (9)$$

where $z = \{z_1, \ldots, z_D\}$, $v = \{v_1, \ldots, v_D\}$, $\alpha = \{\alpha^{(0)}, \ldots, \alpha^{(L)}\}$, and the variables with subscript$\backslash dn$ represent sets of topics, categories, and counts, except for the

n-th count of the d-th dataset. To estimate Dirichlet hyperparameters α, we can apply Minka's fixed-point iteration [1,16]. Hyperparameters $\alpha = \{\alpha^{(0)}, \ldots, \alpha^{(L)}\}$ are updated after each cycle of the collapsed Gibbs sampling by

$$\alpha_k^{(0)} = \alpha_k^{(0)} \frac{\sum_{d=1}^{D} \Psi(N_{dk}^{(0)} + \alpha_k^{(0)}) - D\Psi(\alpha_k^{(0)})}{\sum_{d=1}^{D} \Psi(N_d + \alpha_{\bullet}^{(0)}) - D\Psi(\alpha_{\bullet}^{(0)})}, \tag{10}$$

$$\alpha_{kk'w}^{(l)} = \alpha_{kk'w}^{(l)} \frac{\sum_{d=1}^{D} \Psi(N_{dkk'w}^{(l-1:l)} + \alpha_{kk'w}^{(l)}) - D\Psi(\alpha_{kk'w}^{(l)})}{\sum_{d=1}^{D} \Psi(N_{dk}^{(l-1)} + \alpha_{k\bullet\bullet}^{(l)}) - D\Psi(\alpha_{k\bullet\bullet}^{(l)})}, \tag{11}$$

$$\alpha_w^{(L)} = \alpha_w^{(L)} \frac{\sum_{k=1}^{K_L} \Psi(N_{kw}^{(L-1:L)} + \alpha_w^{(L)}) - K_L\Psi(\alpha_w^{(L)})}{\sum_{k=1}^{K_L} \Psi(N_k^{(L-1)} + \alpha_{\bullet}^{(L)}) - K_L\Psi(\alpha_{\bullet}^{(L)})} \tag{12}$$

where Ψ is a digamma function. Iterating these formulae maximizes the conditional likelihoods in Eqs. (2), (3), and (4), respectively. Thus, latent topics and hyperparameters are updated alternately. The algorithm of the collapsed Gibbs sampling for the CMA model is described in Algorithm 1. From the sample of topics obtained by this algorithm, we can also estimate the multinomial probabilities by

$$\theta_{dk}^{(0)} = \frac{N_{dk}^{(0)} + \alpha_k^{(0)}}{N_d + \alpha_{\bullet}^{(0)}}, \tag{13}$$

$$\theta_{dkk'w}^{(l)} = \frac{N_{dkk'w}^{(l-1:l)} + \alpha_{kk'w}^{(l)}}{N_{dk}^{(l-1)} + \alpha_{k\bullet\bullet}^{(l)}}, \tag{14}$$

$$\theta_{kw}^{(L)} = \frac{N_{kw}^{(L-1:L)} + \alpha_w^{(L)}}{N_k^{(L-1)} + \alpha_{\bullet}^{(L)}}. \tag{15}$$

4   Application to Step-Count Dataset

We applied the proposed model to the step-count datasets recorded by the activity monitors. The average number of steps per day in Japan has been gradually decreasing over the long term, prompting the Japanese government to set the target number of steps for men and women to promote physical activity. The amount of physical activity is certainly important for health promotion, but other lifestyle factors also need to be considered [13]. Nowadays, smart devices attachable to humans, such as activity monitors, are developed to monitor life and determine an individual lifecycle of physical activities from their records. Latent class analysis (LCA) has been used to classify weekly patterns of moderate-to-vigorous physical activity from accelerometer records, and their association with risk factors related to metabolic syndrome has been investigated [14]. LCA and cluster analysis are frequently used approaches in healthcare research on physical activities [8]. However, they classify the amount and patterns of exercise simultaneously, making it difficult to decompose their respective effects on health factors.

Algorithm 1. Collapsed Gibbs sampling for CMA model

Input: $v = \{v_{dn} = (v_{dn}^{(1)}, \ldots, v_{dn}^{(L)}); \ n = 1, \ldots, N_d, \ d = 1, \ldots, D\}$.
Output: $z = \{z_{dn} = (z_{dn}^{(0)}, \ldots, z_{dn}^{(L-1)}); \ n = 1, \ldots, N_d, \ d = 1, \ldots, D\}$,
$\quad \alpha = \{\alpha^{(0)}, \ldots, \alpha^{(L)}\}$.

1: Initialize z, α.
2: Compute $N_{dk}^{(l-1)}, N_{dkk'w}^{(l-1:l)}, N_k^{(L-1)}, N_{kw}^{(L-1:L)}$ by Eqs. (5),(6),(7), and (8).
3: **for** $t = 1, \ldots, T$ **do**
4:     **for** $d = 1, \ldots, D$ **do**
5:        **for** $n = 1, \ldots, N_d$ **do**
6:          Subtract 1 from $N_{dz_{dn}^{(l-1)}}^{(l-1)}, N_{dz_{dn}^{(l-1)}z_{dn}^{(l)}v_{dn}^{(l)}}^{(l-1:l)}$ for each $l = 1, \ldots, L-1$ and
$\qquad N_{z_{dn}^{(L-1)}}^{(L-1)}, N_{z_{dn}^{(L-1)}v_{dn}^{(L)}}^{(L-1:L)}$.
7:          Sample $z_{dn} = (z_{dn}^{(0)}, \ldots, z_{dn}^{(L-1)})$ from its conditional distribution
$\qquad p(z_{dn}|z_{\backslash dn}, v, \alpha)$ given by Eq. (9).
8:          Add 1 to $N_{dz_{dn}^{(l-1)}}^{(l-1)}, N_{dz_{dn}^{(l-1)}z_{dn}^{(l)}v_{dn}^{(l)}}^{(l-1:l)}$ for each $l = 1, \ldots, L-1$ and
$\qquad N_{z_{dn}^{(L-1)}}^{(L-1)}, N_{z_{dn}^{(L-1)}v_{dn}^{(L)}}^{(L-1:L)}$.
9:        **end for**
10:     **end for**
11:     Update $\alpha = \{\alpha^{(0)}, \ldots, \alpha^{(L)}\}$ by Eqs. (10),(11), and (12).
12: **end for**

Therefore, we applied the CMA model to step-count datasets to extract in-week and in-day patterns of ambulatory activities. We estimated two-layered topics with super-topics as weekly activity patterns and sub-topics as daily activity patterns. Using these results, daily and weekly activities were decomposed into the extracted patterns, and the differences in the effects on body weight reduction from the same amounts of activity patterns were discussed.

4.1 Dataset

The step-count dataset in this analysis was recorded by TANITA AM150 and AM160 activity monitors from April 2013 to March 2014. We aggregated hourly step-count records into one-week units from 00:00 on Mondays to 23:59 on Sundays. One-week units with extreme counts over 8,000 steps in an hour or over 50,000 steps in a day were omitted from our analysis. As a result, a dataset of 7,319 weeks from 728 people in Japan whose ages range from 17 to 89 years was obtained. All the missing data where the activity monitor did not work were filled by zero counts.

4.2 Classification of Ambulatory Activities

We applied the proposed model to the step-count dataset described in the previous subsection. The hourly step counts for each one-week unit have $L = 2$ level categories, $W_1 = 7$ days of the week, and $W_2 = 24$ h of the day. Hence, we allocate two-level topic layers and extract weekly activity patterns as super-topics and daily activity patterns as sub-topics. When applying topic models, we need to select the number of topics. We gradually increased the number of topics as long as the extracted topics were easy to interpret and chose $K_0 = 6$ super-topics as weekly patterns and $K_1 = 4$ sub-topics as daily patterns. We ran four MCMC chains in parallel and obtained the sample of the 2,000th iteration from each chain, and then calculated their averages as estimates of latent topics and model parameters.

First, the extracted $K_1 = 4$ sub-topics are shown in Fig. 3 as probability masses of multinomial distributions over the hours of the day given by Eq. (15). The active hours of the four daily patterns are well-separated, and we named the four patterns commuting activity, daytime activity, nighttime activity, and early-bird activity according to their active hours. If we increase the number of sub-topics to $K_1 = 5$, the active hours of the five extracted patterns are almost even partitions of the time from early morning to midnight, which is less interpretable in view of daily lifestyles than the $K_1 = 4$ extracted daily patterns.

Next, Fig. 4 shows the extracted weekly patterns as Dirichlet hyperparameters of $K_0 = 6$ super-topics. Here, we do not show the multinomial probabilities given by Eq. (14) because they differ between one-week units. In Fig. 4(a)–(d), the two daily patterns, commuting and daytime activities, are further divided into weekday (Monday through Friday) and weekend (Saturday and Sunday) activities, respectively. Figure 4(e)–(f) shows nighttime and early-bird activities over a week. Therefore, we named the six weekly patterns as follows: weekday commuting activity, weekend commuting activity, weekday daytime activity, weekend daytime activity, nighttime activity, and early-bird activity. Note that there remain some overlaps among the former four weekly patterns. For example, the values of Dirichlet hyperparameters for the commuting and daytime activities during weekdays are slightly high in the weekend commuting activity, while those for the commuting activity during weekdays are slightly high in the weekend daytime activity. If we increase the number of super-topics from $K_0 = 6$, the weekend activity patterns are further divided into Saturday's activity and Sunday's activity.

For comparison, we show the six weekly patterns extracted by LDA in Fig. 5. Since LDA extracts weekly patterns directly from the hourly counts over a week, each extracted pattern has slightly different daily patterns among the days of the week. However, there are no weekly patterns whose activities are concentrated on weekdays or weekends as in Fig. 4(a)–(d). Even if we increase the number of topics in LDA, the weekly active days are hardly divided while the daily active hours are further divided by the increased topics. Owing to the hierarchical structure of categories, the CMA model can extract and interpret the weekly and daily activity patterns hierarchically.

4.3   Comparison with Tensor Train Decomposition

Our hierarchical topic model for tensor data is similar to another method for tensor data called tensor train decomposition (TTD) [17].

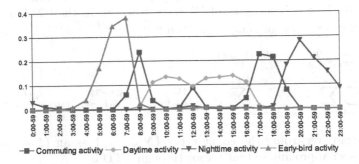

Fig. 3. Daily patterns of ambulatory activity extracted by the CMA model

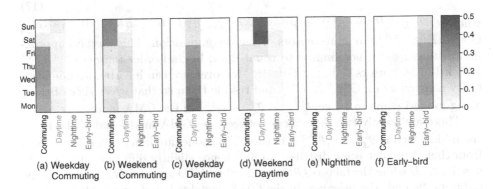

(a) Weekday Commuting (b) Weekend Commuting (c) Weekday Daytime (d) Weekend Daytime (e) Nighttime (f) Early–bird

Fig. 4. Estimated Dirichlet hyperparameters as weekly activity patterns

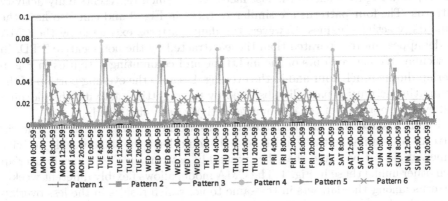

Fig. 5. Weekly patterns of ambulatory activity extracted by LDA

To observe the similarity, we define $N_{dw_1 \cdots w_L}$ as the counts in the d-th dataset belonging to the w_l-th category of $\mathcal{W}_l$ in each mode. In the CMA model, the conditional expectation of $N_{dw_1 \cdots w_L}$, given $\theta_d^{(0)}, \ldots, \theta_d^{(L-1)}, \theta^{(L)}$, is obtained by

$$
\begin{aligned}
& E[N_{dw_1 \cdots w_L} | \theta_d^{(0)}, \ldots, \theta_d^{(L-1)}, \theta^{(L)}] \\
&= N_d \sum_{k_0=1}^{K_0} \sum_{k_1=1}^{K_1} \cdots \sum_{k_{L-2}=1}^{K_{L-2}} \sum_{k_{L-1}=1}^{K_{L-1}} \theta_{dk_0}^{(0)} \theta_{dk_0 k_1 w_1}^{(1)} \cdots \theta_{dk_{L-2} k_{L-1} w_{L-1}}^{(L-1)} \theta_{k_{L-1} w_L}^{(L)}. \quad (16)
\end{aligned}
$$

In contrast, regarding the count array $\{N_{dw_1 \cdots w_L}\}$ as an $(L+1)$-st order tensor, we can approximate that tensor through TTD by

$$
N_{dw_1 \cdots w_L} \simeq N_d \sum_{k_0=1}^{K_0} \sum_{k_1=1}^{K_1} \cdots \sum_{k_{L-2}=1}^{K_{L-2}} \sum_{k_{L-1}=1}^{K_{L-1}} G_{dk_0}^{(0)} G_{k_0 w_1 k_1}^{(1)} \cdots G_{k_{L-2} w_{L-1} k_{L-1}}^{(L-1)} G_{k_{L-1} w_L}^{(L)}
$$
$$(17)$$

where $G^{(0)} \in \mathbb{R}^{D \times K_0}, G^{(1)} \in \mathbb{R}^{K_0 \times W_1 \times K_1}, \ldots, G^{(L-1)} \in \mathbb{R}^{K_{L-2} \times W_{L-1} \times K_{L-1}}$, $G^{(L)} \in \mathbb{R}^{K_{L-1} \times W_L}$ are core tensors of the decomposition. This method approximates tensor data, not limited to count data, by 2nd-order tensors $G^{(0)}, G^{(L)}$ and 3rd-order tensors $G^{(1)}, \ldots, G^{(L-1)}$. Moreover, we can incorporate nonnegativity constraints on $G^{(0)}, \ldots, G^{(L)}$ and rescale them so that every slice of them along mode-1 sums to 1 like $\theta_d^{(0)}, \ldots, \theta_d^{(L-1)}, \theta^{(L)}$ in the CMA model.

There are mainly two differences between the CMA model and TTD. First, the middle factors $\theta_d^{(1)}, \ldots, \theta_d^{(L-1)}$ in the CMA model are independently sampled from their Dirichlet prior distribution and hence slightly differ among datasets $d = 1, \ldots, D$ while the factors $G^{(1)}, \ldots, G^{(L)}$ in TTD are common among all the datasets. Second, the inference in the CMA model is based on MCMC methods while that in TTD is based on alternating least squares algorithms.

For comparison with the previous results obtained by the CMA model, we apply the nonnegative TTD [7] to the same step-count data with $L = 2$, $K_0 = 6$, and $K_1 = 4$. Figure 6 shows the last factor $G^{(2)}$, which represents daily activity patterns. The four patterns are similar to those in Fig. 3 and can also have the same daily activity names. However, the daily patterns extracted by the CMA model appear more separated than those extracted by the nonnegative TTD. In particular, the active hours of the nighttime and commuting activities in Fig. 6 overlap during lunch hour and in the evening. Because the evening portion of the commuting activity is partially absorbed by the nighttime activity, the morning and evening portions of the commuting activity are unbalanced.

Figure 7 shows the middle factor $G^{(1)}$ as weekly activity patterns. The six patterns are similar to those in Fig. 4 and should have the same weekly activity names. It should be noted that the six weekly patterns in Fig. 7 hardly overlap each other unlike those in Fig. 4. The CMA model allows flexibly different weekly patterns among the datasets in the same topic, which results in the less overlap among the daily patterns in Fig. 3.

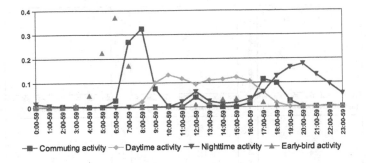

-■- Commuting activity   -◆- Daytime activity   -▼- Nighttime activity   -▲- Early-bird activity

Fig. 6. Daily patterns of ambulatory activity extracted by the nonnegative TTD

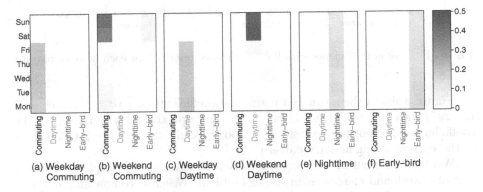

(a) Weekday   (b) Weekend   (c) Weekday   (d) Weekend   (e) Nighttime   (f) Early–bird
Commuting   Commuting   Daytime   Daytime

Fig. 7. Weekly patterns of ambulatory activity extracted by the nonnegative TTD

4.4   Effects of Weekly Activity Patterns on Body Weight Changes

Using the extracted weekly activity patterns, we investigated the effect of ambulatory activities on body weight changes. We obtained daily body weight records measured over the same weeks as the step-count data from 254 of the 728 people studied in the previous analysis.

We applied the following linear mixed model to estimate the effect of the number of steps in each weekly activity pattern on body weight changes:

$$x_{dw} = \sum_{k=1}^{6} \beta_{kw} n_{dk} + \xi_{i_d w} + \varepsilon_{dw}, \tag{18}$$

where x_{dw} is the percentage of body weight change from 00:00 on Monday to 24:00 on the w-th day in the d-th dataset; n_{dk} is the step count, divided by 100,000, of the k-th weekly activity pattern in the d-th dataset; β_{kw} is a regression coefficient for the step count of the k-th weekly activity pattern; $\xi_{i_d w}$ is a random effect for the i_d-th individual to be measured in the d-th dataset; and ε_{dw} is the residual term. Because body weight is typically not measured at 24:00, we interpolated hourly body weights to obtain x_{dw} by applying a simple state space model called the local level model [4].

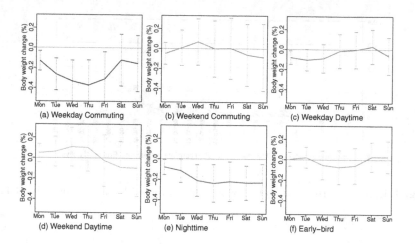

Fig. 8. Regression coefficients with 95% confidence intervals for each activity pattern

Figure 8 shows the estimated regression coefficients β_{kw} ($k = 1, \ldots, 6$, $w = 1, \ldots, 7$) with 95% confidence intervals for the respective activity patterns. The coefficients represent the percentages of body weight changes per 100,000 steps of the corresponding activity pattern.

Weekday commuting and daytime activities reduce body weight in the first half of a week and increase it in the second half. Weekend commuting and daytime activities have the opposite effects, but they are not significant. Nighttime activity significantly reduces body weight throughout a week, but early-bird activity does not.

5   Concluding Remarks

We proposed a hierarchical topic model, called the CMA model, to extract common patterns from count-valued tensor data. Despite the multilayer modeling, sharing common topics in each mode provides simplicity and interpretability of the extracted topics. Latent topics, their membership, and hyperparameters can be estimated via collapsed Gibbs sampling.

We applied the proposed model to activity monitor records and extracted weekly and daily patterns of ambulatory activities. The extracted patterns seem to reflect the weekly and daily activities of typical lifestyles. Furthermore, we investigated the effect of activity in each weekly pattern on body weight changes within a week. Some weekly patterns have significant effects on body weight reduction during their active days. The weekly and daily activity patterns should also be related to the dynamics of blood pressure and blood-sugar levels, which will be discussed in future works.

The CMA model is closely related to the TTD but provides patterns that are slightly different from those provided by the nonnegative TTD. The topic models whose structures correspond to other types of tensor decomposition and tensor networks may also be constructed in a similar manner.

Acknowledgement. This study was conducted as part of the "Research and Development on Utilization of Fundamental Technologies for Social Big Data" (178A04) project of NICT (National Institute of Information and Communication Technology).

References

1. Asuncion, A., Welling, M., Smyth, P., Teh, Y.W.: On smoothing and inference for topic models. In: Proceedings of the 25th Conference on Uncertainty in Artificial Intelligence, pp. 27–34 (2009)
2. Blei, D.M., Jordan, M.I.: Modeling annotated data. In: Proceedings of the 26th Annual ACM SIGIR Conference on Research and Development in Information Retrieval, pp. 127–134 (2003)
3. Blei, D.M., Ng, A.Y., Jordan, M.I.: Latent Dirichlet allocation. J. Mach. Learn. Res. **3**, 993–1022 (2003)
4. Durbin, J., Koopman, S.J.: Time Series Analysis by State Space Methods, 2nd edn. Oxford University Press, Oxford (2012)
5. Iwata, T., Sawada, H.: Topic model for analyzing purchase data with price information. Data Min. Knowl. Disc. **26**, 559–573 (2013)
6. Iwata, T., Watanabe, S., Yamada, T., Ueda, N.: Topic tracking model for analyzing consumer purchase behavior. In: Proceedings of the 21st International Joint Conference on Artificial Intelligence, pp. 1427–1432 (2010)
7. Lee, N., Phan, A.-H., Cong, F., Cichocki, A.: Nonnegative tensor train decompositions for multi-domain feature extraction and clustering. In: Hirose, A., Ozawa, S., Doya, K., Ikeda, K., Lee, M., Liu, D. (eds.) ICONIP 2016. LNCS, vol. 9949, pp. 87–95. Springer, Cham (2016). https://doi.org/10.1007/978-3-319-46675-0_10
8. Leech, R.M., McNaughton, S.A., Timperio, A.: The clustering of diet, physical activity and sedentary behavior in children and adolescents: a review. Int. J. Behav. Nutr. Phys. Act. **11**(4), 1–9 (2016). https://doi.org/10.1186/1479-5868-11-4
9. Li, W., McCallum, A.: Pachinko allocation: scalable mixture models of topic correlations. In: Proceedings of the 23rd International Conference on Machine Learning, pp. 577–584 (2006)
10. Lu, H., Wei, C., Hsiao, F.: Modeling healthcare data using multiple-channel latent Dirichlet allocation. J. Biomed. Inform. **60**, 210–223 (2016)
11. Liu, J.S.: The collapsed Gibbs sampler in Bayesian computations with applications to a gene regulation problem. J. Am. Stat. Assoc. **89**, 958–966 (1994)
12. Masada, T., Takasu, A.: A topic model for traffic speed data analysis. In: Ali, M., Pan, J.-S., Chen, S.-M., Horng, M.-F. (eds.) IEA/AIE 2014. LNCS (LNAI), vol. 8482, pp. 68–77. Springer, Cham (2014). https://doi.org/10.1007/978-3-319-07467-2_8
13. McAloney, K., Graham, H., Law, C., Platt, L.: A scoping review of statistical approaches to the analysis of multiple health-related behaviours. Prev. Med. **56**, 356–371 (2013). https://doi.org/10.1016/j.ypmed.2013.03.002
14. Metzger, J.S., Catellier, D.J., Evenson, K.R., Treuth, M.S., Rosamond, W.D., Siega-Riz, A.M.: Associations between patterns of objectively measured physical activity and risk factors for the metabolic syndrome. Am. J. Health Promot. **24**(3), 161–169 (2010). https://doi.org/10.4278/ajhp.08051151

15. Mimno, D., Wallach, H.M., Naradowsky, J., Smith, D.A., McCallum, A.: Polylingual topic models. In: Proceedings of the 2009 Conference on Empirical Methods in Natural Language Processing, pp. 880–889 (2009)
16. Minka, T.: Estimating a Dirichlet distribution (2000)
17. Oseledets, I.V.: Tensor-train decomposition. SIAM J. Sci. Comput. **33**(5), 2295–2317 (2011). https://doi.org/10.1137/090752286

Convolutional Neural Network to Detect Deep Low-Frequency Tremors from Seismic Waveform Images

Ryosuke Kaneko[1,2], Hiromichi Nagao[1,2(✉)], Shin-ichi Ito[1,2], Kazushige Obara[2], and Hiroshi Tsuruoka[2]

[1] Graduate School of Information Science and Technology, The University of Tokyo, Tokyo, Japan
`ryosuke-kaneko@g.ecc.u-tokyo.ac.jp`, `nagaoh@eri.u-tokyo.ac.jp`
[2] Earthquake Research Institute, The University of Tokyo, Tokyo, Japan

Abstract. The installation of dense seismometer arrays in Japan approximately 20 years ago has led to the discovery of deep low-frequency tremors, which are oscillations clearly different from ordinary earthquakes. As such tremors may be related to large earthquakes, it is an important issue in seismology to investigate tremors that occurred before establishing dense seismometer arrays. We use deep learning aiming to detect evidence of tremors from past seismic data of more than 50 years ago, when seismic waveforms were printed on paper. First, we construct a convolutional neural network (CNN) based on the ResNet architecture to extract tremors from seismic waveform images. Experiments applying the CNN to synthetic images generated according to seismograph paper records show that the trained model can correctly determine the presence of tremors in the seismic waveforms. In addition, the gradient-weighted class activation mapping clearly indicates the tremor location on each image. Thus, the proposed CNN has a strong potential for detecting tremors on numerous paper records, which can enable to deepen the understanding of the relations between tremors and earthquakes.

Keywords: Deep low-frequency tremor · Convolutional neural network · ResNet · Gradient-weighted class activation mapping

1 Introduction

The surface of the Earth comprises 15 tectonic plates, with each plate individually moving several centimeters per year. Seismic phenomena often occur at the boundary where a plate subducts beneath another one in the subduction zone by releasing strains accumulated over a long time. Figure 1 shows the schematic of a subduction zone, in which an oceanic plate is subducting beneath a continental plate. For example, the Philippine Sea Plate is subducting beneath the Eurasian Plate around southwest Japan, forming a subduction zone called the

M. Gupta and G. Ramakrishnan (Eds.): PAKDD 2021 Workshops, LNAI 12705, pp. 31–43, 2021.
https://doi.org/10.1007/978-3-030-75015-2_4

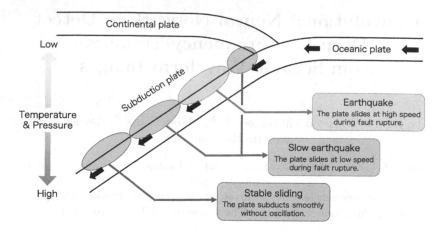

Fig. 1. Schematic of subduction zone. The ellipses represent the locations of seismic phenomena occurrence.

Nankai Trough. In a subduction zone, various seismic phenomena occur depending on the magnitude of plate friction: ordinary earthquakes (high friction), slow earthquakes (low friction), and stable sliding (no friction). The friction generally increases as the temperature decreases and the pressure increases. As both temperature and pressure increase with depth, the friction according to depth presents a complicated relation. This relation might lead to different locations of seismic phenomena occurrence, as shown in Fig. 1. Ordinary earthquakes usually occur when the strains accumulated over a long time are released instantaneously at a shallow depth, where the friction is very high. At the beginning of the 21st century, new phenomena called slow earthquakes originated from subduction zones were discovered [8]. Owing to the lower friction, slow earthquakes repeat with much shorter intervals and have smaller magnitudes than ordinary earthquakes.

Hi-net [9,12], the seismometer network installed by NIED and operating since 1996, led to the discovery of deep low-frequency tremors, which are categorized as slow earthquakes. The Hi-net seismometers are more sensitive and densely located than conventional seismometer networks. Thus, they allow to observe weak oscillations that were previously unobservable and correlatively analyze records between neighboring seismometers. The envelope correlation method [8] successfully extracted evidence of tremors from Hi-net data, which has been recognized as the first discovery of tremors in the world. Such tremors are weak oscillations that occur in a deeper area than ordinary earthquakes. These tremors have an approximate dominant frequency of 2–8 Hz, last from several hours to several days, and have magnitudes below 1.3 according to the tremor catalog published by NIED [4,11]. Thus, tremors cannot be perceived on the ground. According to the NIED catalog, more than 30,000 tremors have been observed in southwest Japan from January 2001 to April 2019, indicating their high frequency. Tremors have been detected in various subduction zones worldwide after the first discovery (*e.g.*, [5,15]), and they represent a research hotspot in seismology, as many studies have been indicated the relations between tremors and large earthquakes (*e.g.*, [10]). In fact, seismologists expect tremors to provide clues to predict large earthquakes and understand their mechanisms. Currently, only digital data from the last 20 years are available for studying tremors. Considering that megathrust earthquakes have periodically occurred in the Nankai Trough over intervals of 100–200 years, it is important to analyze tremor occurrences in southwest Japan over a longer period. Before the seismic records were available in digital format, seismometers continuously recorded waveforms with a pen on drum-rolled papers.

In this study, we aimed to detect tremors from seismograph paper records by using a convolutional neural network (CNN), a deep-learning method that has shown high performance for image recognition. A CNN can automatically tune its internal parameters by learning the characteristics of tremors from input images without requiring prior knowledge of tremors or manually adjusting the parameters. Training a CNN from scratch with real data polluted by a variety of noises may hinder the model construction and hyperparameter tuning. Thus, we conducted numerical experiments to construct a CNN and train it with synthetic images similar to the seismograph paper records. These experiments aim to obtain clues to improve the model and provide pretrained models for subsequent fine-tuning, which is a popular learning method to improve the model performance and learning efficiency.

2   Seismographs

Modern seismic research based on observational data strongly relies on digital records. In Hi-net, more than 1000 seismometers installed in Japan Islands continuously observe the velocity of the ground at a sampling rate 100 Hz, and NIED collects and publishes the corresponding digital data in real time. Multivariate time-series analyses on digital seismic data considering spatial correlations enable to increase the signal-to-noise ratio of detected phenomena or eventually uncover unknown phenomena such as slow earthquakes and tremors.

Past seismometers used over 50 years ago drew waveforms directly on paper. Considering the time interval between megathrust earthquakes in the subduction zones, paper records are a valuable source for research on slow earthquakes (*e.g.*, [2]). Figure 2 shows an example of a seismograph paper record. The daily records are drawn on a single piece of paper, in which each time series is drawn horizontally and contains approximately 2.5 min of data. Besides seismic waveforms, the records contain pulses that indicate a time stamp every second. The average image size of a paper record excluding its margins is approximately 7000 × 7000 pixels.

The digitization of paper records by tracing the waveforms is effective for investigating large earthquakes because such waveforms are extractable even from overlapping time series given the low frequencies and large amplitudes of earthquakes (*e.g.*, [6]). In contrast, tremors generally have smaller amplitudes

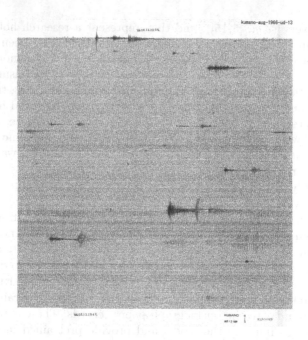

Fig. 2. Seismograph paper record from August 13, 1966, 19:47 to August 14, 1966, 20:05 in Kumano, Japan [14].

and higher frequencies than large earthquakes. Consequently, their digitization is difficult, especially for overlapping waveforms. Therefore, CNN-based image recognition is a promising alternative for analyzing tremors compared to individual waveform extraction through digitization.

Both digital and analog seismic data include a wide variety of phenomena such as earthquakes, tremors, pulsations excited by oceanic waves, teleseisms (distant earthquakes), oscillations due to meteorological events, and noise. Thus, identifying tremors from seismic data becomes difficult when large earthquakes or other signals in similar spectra pollute the measurements.

3   Methods

To detect tremors by applying CNNs to real data, we conducted numerical experiments based on synthetic images to obtain a trained model that determines correctly the presence of tremors in an input image. In this section, we detail these experiments, including image synthetization and CNN construction.

3.1   Generation of Synthetic Images

We generate synthetic images based on seismograph paper records from Kumano, Japan (Fig. 2). Figure 3 shows examples of synthetic images. Each image is

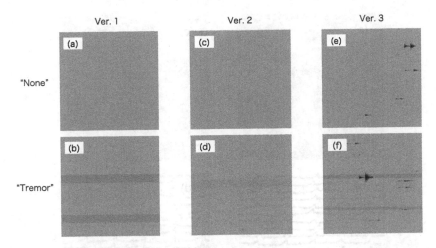

Fig. 3. Representative examples of synthetic images. Details of each image are available on the text and Table 1.

Table 1. Versions of synthetic images. The checkmarks represent the waveforms contained in each version. The values next to the checkmarks indicate the ratio of the waveform amplitude with respect to Ver. 1.

Waveforms	Ver. 1	Ver. 2	Ver. 3
Observation noise (Gaussian noise)	✓	✓	✓
Time stamps (pulses every second)	✓	✓	✓
Pulsations (Gaussian noise + bandpass filter)	✓ (1.0)	✓ (2.0)	✓ (2.0)
Tremors (Gaussian noise + bandpass filter)	✓ (1.0)	✓ (0.5)	✓ (1.0)
Earthquakes (P waves & S waves)			✓

7000×7000 pixels and corresponds to a daily record that contains 576 time series of 2.5 min vertically stacked. The synthetic images correspond to one of three versions, Ver. 1, Ver. 2, or Ver. 3, according to the included types of signals and noise listed in Table 1. For each version, 100 images were generated without tremors (labeled "none"), and 100 images were generated with tremors (labeled "tremor"). Ver. 1 images (Figs. 3(a) and 3(b)) contain observation noise, time stamps, and pulsations, as well as tremors only for "tremor" images. Ver. 1 represents a simple case that allows straightforward tremor detection. In Ver. 2 images (Figs. 3(c) and 3(d)), the tremors are smaller and the pulsations are larger than those in Ver. 1 images. In Ver. 3 images (Figs. 3(e) and 3(f)), we change the amplitudes of tremors and pulsations and insert earthquakes. Figure 4 confirms that Ver. 3 images suitably resemble the paper records regarding both their overview and details.

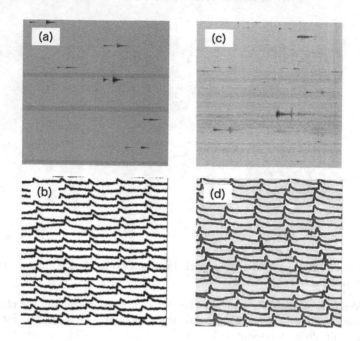

Fig. 4. (a) Ver. 3 "tremor" image. (b) Magnified view of (a). (c) Image of paper record shown in Fig. 2. (d) Magnified view of (c).

3.2 Preprocessing of Synthetic Images

For preprocessing, we divide each image vertically into five strips of 7000×1400 pixels and then resize each strip to 2000×400 pixels. The vertical division allows to easily distinguish between tremors and noise. As a tremor usually lasts several hours, the five strips must include its evidence, unlike temporary noise. This feature is useful to decide the presence of tremors on the images. For example, consider five strips extracted from an image with unknown label, either "none" or "tremor." If a CNN assigns "none" to four of the five strips and "tremor" to the remaining strip, we can assume that the image corresponds to "none", and the misjudgment for the last strip is due to noise. The reduced resolution after resizing aims to reduce the number of model parameters and consequently the computational cost. After preprocessing, each version comprises 500 "none" images and 500 "tremor" images. For each version, we use 800 images for training and the remaining 200 for validation, as detailed in the next subsection.

3.3 CNN for Tremor Detection

The CNN is a representative deep learning method that has exhibited high performance in tasks such as image recognition and handwriting recognition. A CNN has two distinctive layers, the convolutional layer and the pooling layer, which mathematically describe the function of human visual cells. These layers

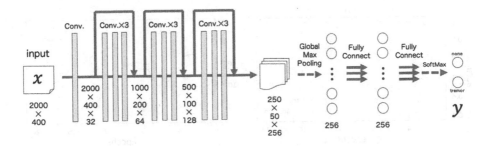

Fig. 5. Architecture of proposed CNN.

allow the CNN to extract features from input images. In recent years, CNNs have begun to be used in seismology for tasks such as detection or discrimination of seismic phenomena (*e.g.*, [7,13]).

Below, we define the key terms and formulas of CNNs. Let x be an input image for a CNN. Image x has a true label ℓ in set S of labels. Let $\mathcal{M}$ be a CNN with internal parameters $\boldsymbol{\theta}$. For an input x, CNN $\mathcal{M} = \mathcal{M}(\boldsymbol{\theta})$ provides predictions $\boldsymbol{y} = (p_\ell)_{\ell \in S}$, where p_ℓ is the predicted probability that the true label of x is ℓ. In a CNN, the loss is a function that expresses its performance. The loss returns a non-negative real number for a pair $(\ell, \boldsymbol{y})$ of the same x. If the loss value for x is close to 0, the prediction performance for x is favorable. On the other hand, the accuracy is the agreement rate between true label ℓ and predicted label $\hat{\ell} = \arg\max_{\ell \in S} p_\ell$ for the inputs. Training is the process to optimize parameters $\boldsymbol{\theta}$ by minimizing the loss value for inputs with known labels. Then, validation calculates the loss and accuracy at fixed parameters $\boldsymbol{\theta}$ for inputs with known labels but not used for training. Thus, validation evaluates the predictive performance of the CNN for previously unseen inputs. Finally, test performs prediction on inputs with unknown labels.

We built the proposed CNN by incorporating the residual connections used in the ResNet [1]. Figure 5 shows the architecture of the proposed CNN, which establishes a binary classifier that determines whether the input image has label "none" or "tremor." The CNN output has the form $\boldsymbol{y} = (p_{\text{"none"}}, p_{\text{"tremor"}})$ corresponding to $S = \{\text{"none"}, \text{"tremor"}\}$. We used the categorical cross-entropy as the loss function and optimized it using the Adam method [3] with a batch size of 16 in this study.

4   Results

Figure 6 shows the accuracy and loss throughout learning of Ver. 1 images. An epoch (horizontal axis) indicates learning iterations, that is, a set of training using all the training images and validation using all the validation images. Although the validation accuracy does not improve in the initial learning stage, it rapidly increases after 13 epochs to reach almost 1.0, possibly after the model parameters leave a local optimum. Figure 7 shows the gradient-weighted class

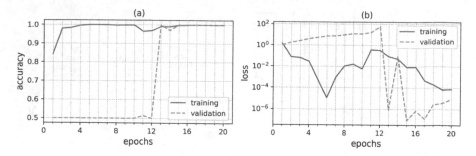

Fig. 6. (a) Accuracy and (b) loss for Ver. 1 images according to number of training epochs.

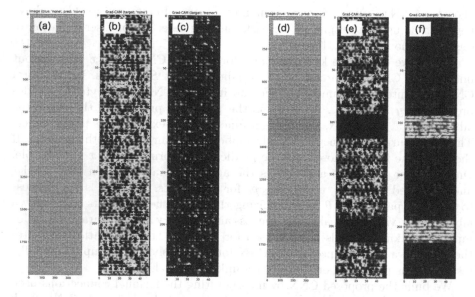

Fig. 7. (a) Ver. 1 "none" image and corresponding Grad-CAMs for prediction of labels (b) "none" and (c) "tremor." (d) Ver. 1 "tremor" image and corresponding Grad-CAMs for prediction of labels (e) "none" and (f) "tremor."

activation maps (Grad-CAMs) [16] of Ver. 1 images. Each Grad-CAM indicates the image regions that influence the prediction calculation, with red (light) indicating the highest influence and blue (dark) indicating no influence. For prediction of "none", the response is uniform on the entire image, except for the areas containing tremors (Figs. 7(b) and 7(e)). In Fig. 7(f), remarkable responses clearly appear in areas containing tremors. Therefore, the model correctly detects tremors and correctly identifies the "tremor" image.

Figures 8 and 9 show the learning performance and Grad-CAMs for Ver. 2 images, respectively. Although the tremors in Ver. 2 are smaller than those in Ver. 1, model training and tremor detection are successful.

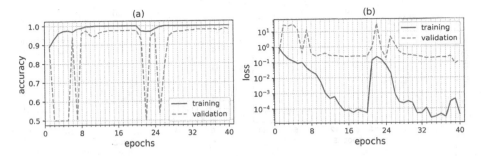

Fig. 8. (a) Accuracy and (b) loss for Ver. 2 images according to number of training epochs.

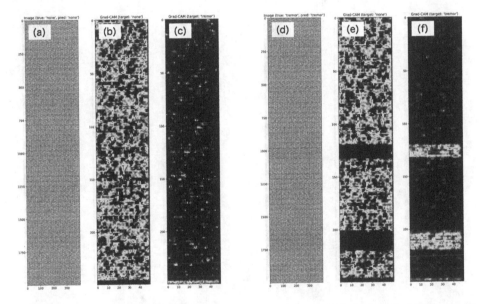

Fig. 9. (a) Ver. 2 "none" image and corresponding Grad-CAMs for prediction of labels (b) "none" and (c) "tremor." (d) Ver. 2 "tremor" image and corresponding Grad-CAMs for prediction of labels (e) "none" and (f) "tremor."

Figures 10 and 11 show the learning performance and Grad-CAMs for Ver. 3 images, respectively. In Ver. 3 images, the Grad-CAM responses to earthquake waveforms are notable. For small earthquake waveforms, the Grad-CAMs do not show any remarkable response. This result may be due to the input image shrinking as it passes through the convolutional layers, and small waveforms eventually disappear during shrinking. On the other hand, in Figs. 11(b) and 11(e), the large earthquake waveforms cause the most influential responses. This result suggests that the CNN can distinguish earthquake waveforms from tremors, observation noise, and pulsations. Moreover, the proposed CNN may be able to achieve multinomial classification of labels such as "none," "tremor," "earthquake," and

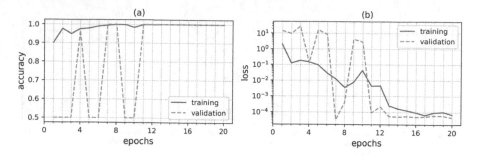

Fig. 10. (a) Accuracy and (b) loss for Ver. 3 images according to number of training epochs.

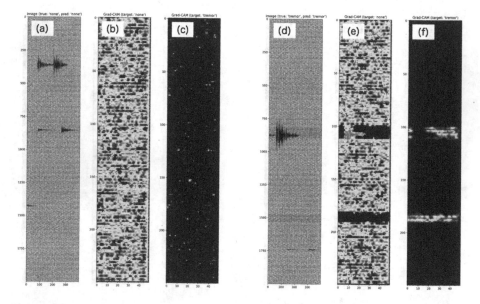

Fig. 11. (a) Ver. 3 "none" image and corresponding Grad-CAMs for prediction of labels (b) "none" and (c) "tremor." (d) Ver. 3 "tremor" image and corresponding Grad-CAMs for prediction of labels (e) "none" and (f) "tremor."

"both." Figure 11(f) shows tremor responses that do not appear when tremors are masked by earthquake waveforms.

Overall, the tremor waveforms appear to show a band-like pattern in the images. This result may indicate that the CNN discriminates tremors based on a rough view of the image. To discard this possibility, we conducted an additional experiment using monochromatic and dichromatic images, as shown in Fig. 12. The set of monochromatic images consists of 256 images for grayscale values from 0 to 255. The set of dichromatic images consists of 200 images, with each image containing a monochromatic band-like pattern on a monochromatic background. By providing these images as inputs to the CNN trained on Ver. 3, all the

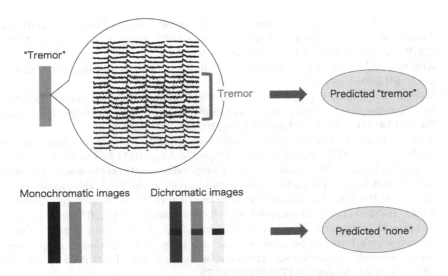

Fig. 12. Proposed CNN trained on Ver. 3 images correctly predicts label "tremor" for "tremor" images and label "none" for all monochromatic and dichromatic images.

predicted labels are "none." Although Ver. 3 "tremor" images seem to be similar to the dichromatic images, the CNN correctly determines the presence of tremors in Ver. 3 images. This result indicates that the CNN appropriately learned tremor features.

5   Conclusion

The proposed CNN is expected to effectively detect tremors from seismograph paper records, as verified through numerical experiments. Based on the finding from the experiments, we will conduct CNN training with real data, which contain a wider variety of noises than synthetic images. To improve the CNN performance, we will also explore persistent parameter tuning and additional data preprocessing methods.

Acknowledgement. This work was supported by JST CREST under Grant Numbers JPMJCR1763 and JPMJCR1761. The key ideas in this study derived from the activities of JSPS KAKENHI Grant-in-Aids for Scientific Research (B) No. 17H01703, 17H01704, 18H03210, Grant-in-Aid for Scientific Research (S) No. 19H05662, Grant-in-Aid for Challenging Research (Exploratory) No. 20K21785, and Earthquake Research Institute Joint Research ERI JURP 2020-A-05.

References

1. He, K., Zhang, X., Ren, S., Sun, J.: Deep residual learning for image recognition. In: 2016 IEEE Conference on Computer Vision and Pattern Recognition (CVPR), pp. 770–778 (2016). https://doi.org/10.1109/CVPR.2016.90

2. Kano, M., Kano, Y.: Possible slow slip event beneath the Kii Peninsula, southwest Japan, inferred from historical tilt records in 1973. Earth Planets Space **71**(95) (2019). https://doi.org/10.1186/s40623-019-1076-9
3. Kingma, D.P., Ba, J.: Adam: a method for stochastic optimization (2017). arXiv:1412.6980 [cs.LG]
4. Maeda, T., Obara, K.: Spatiotemporal distribution of seismic energy radiation from low-frequency tremor in western Shikoku, Japan. J. Geophys. Res.: Solid Earth **114**(B10) (2009). https://doi.org/10.1029/2008JB006043
5. Nadeau, R.M., Dolenc, D.: Nonvolcanic tremors deep beneath the San Andreas Fault. Science **307**(5708), 389 (2005). https://doi.org/10.1126/science.1107142
6. Nagai, R., Kikuchi, M., Yamanaka, Y.: Comparative study on the source processes of recurrent large earthquakes in Sanriku-oki region: the 1968 Tokachi-oki earthquake and the 1994 Sanriku-oki earthquake. Zisin (J. Seismol. Soc. Jpn. 2nd ser.) **54**(2), 267–280 (2001). https://doi.org/10.4294/zisin1948.54.2_267. (in Japanese)
7. Nakano, M., Sugiyama, D., Hori, T., Kuwatani, T., Tsuboi, S.: Discrimination of seismic signals from earthquakes and tectonic tremor by applying a convolutional neural network to running spectral images. Seismol. Res. Lett. **90**(2A), 530–538 (2019). https://doi.org/10.1785/0220180279
8. Obara, K.: Nonvolcanic deep tremor associated with subduction in southwest Japan. Science **296**(5573), 1679–1681 (2002). https://doi.org/10.1126/science.1070378
9. Obara, K., Kasahara, K., Hori, S., Okada, Y.: A densely distributed high-sensitivity seismograph network in Japan: Hi-net by National Research Institute for Earth Science and Disaster Prevention. Rev. Sci. Instrum. **76**(2) (2005). https://doi.org/10.1063/1.1854197
10. Obara, K., Kato, A.: Connecting slow earthquakes to huge earthquakes. Science **353**(6296), 253–257 (2016). https://doi.org/10.1126/science.aaf1512
11. Obara, K., Tanaka, S., Maeda, T., Matsuzawa, T.: Depth-dependent activity of non-volcanic tremor in southwest Japan. Geophys. Res. Lett. **37**(13) (2010). https://doi.org/10.1029/2010GL043679
12. Okada, Y., et al.: Recent progress of seismic observation networks in Japan—Hi-net, F-net, K-NET and KiK-net. Earth Planets Space **56**(8), xv–xxviii (2004). https://doi.org/10.1186/BF03353076
13. Perol, T., Gharbi, M., Denolle, M.: Convolutional neural network for earthquake detection and location. Sci. Adv. **4** (2017). https://doi.org/10.1126/sciadv.1700578
14. Satake, K., Tsuruoka, H., Murotani, S., Tsumura, K.: Analog seismogram archives at the Earthquake Research Institute, the University of Tokyo. Seismol. Res. Lett. **91**(3), 1384–1393 (2020). https://doi.org/10.1785/0220190281
15. Schwartz, S.Y., Rokosky, J.M.: Slow slip events and seismic tremor at circum-Pacific subduction zones. Rev. Geophys. **45**(3) (2007). https://doi.org/10.1029/2006RG000208
16. Selvaraju, R.R., Cogswell, M., Das, A., Vedantam, R., Parikh, D., Batra, D.: Grad-cam: visual explanations from deep networks via gradient-based localization. In: 2017 IEEE International Conference on Computer Vision (ICCV), pp. 618–626 (2017). https://doi.org/10.1109/ICCV.2017.74

Unsupervised Noise Reduction for Nanochannel Measurement Using Noise2Noise Deep Learning

Takayuki Takaai[✉] and Makusu Tsutsui

The Institute of Scientific and Industrial Research, Osaka University, Suita, Japan
takaai@ar.sanken.osaka-u.ac.jp

Abstract. Noise reduction is an important issue in measurement. A difficulty to train a noise reduction model using machine learning is that clean signal on measurement object needed for supervised training is hardly available in most advanced measurement problems. Recently, an unsupervised technique for training a noise reduction model called Noise2Noise has been proposed, and a deep learning model named U-net trained by this technique has demonstrated promising performance in some measurement problems. In this study, we applied this technique to highly noisy signals of electric current waveforms obtained by measuring nanoparticle passages in a multistage narrowing nanochannel. We found that a convolutional AutoEncoder (CAE) was more suitable than the U-net for the noise reduction using the Noise2Noise technique in the nanochannel measurement problem.

Keywords: Nanochannel · Noise reduction · Noise2Noise · Measurement · Deep learning

1 Introduction

Many advanced measurement devices and apparatus used today employ highly elaborated measurement principles and processes under their extreme target conditions such as extremely minute molecules or living bodies, ultimately high resolution images and ultimately short time phenomena. Their measurement objects often include complex structures with many degrees of freedom, considerable fluctuations and noises in both the time and spatial domains. Therefore, enormous volumes of measurement data output from the measurement devices and apparatus often include incomplete information on the measurement objects. Under these circumstances, the necessity of using advanced principles for highly accurate estimation of target phenomena from the measurement information buried in noise has been increasing.

As one example of measurement problems using such advanced measurement devices, we have been tackling effective noise reduction of the electric current waveform representing nanoparticle passage measured by multistage narrowing

M. Gupta and G. Ramakrishnan (Eds.): PAKDD 2021 Workshops, LNAI 12705, pp. 44–56, 2021.
https://doi.org/10.1007/978-3-030-75015-2_5

nanochannels [1]. Ultimate goal of the development of this multistage narrowing nanochannels as a measurement device is the acquisition of information related to various behaviors and properties of the particles such as the masses of the particles and their passage times.

However, the space measured by the multistage narrowing nanochannel and the size of the targeted nanoparticles are extremely tiny, and thus the electric current obtained by this measurement is extremely weak. Therefore, the fluctuation of particles, the thermal measurement noise, and the signal noise caused by electric circuits used in the measurement devices are superimposed to the electric current waveform of the measurement output. Since the generations and mixings of these noises are very complicated nonlinear processes, any appropriate models of these processes based on physical and chemical viewpoints are hardly obtained.

Our primary objective in this study is to develop a technique effectively reducing the noise in the electric current waveform signals without using any physical and chemical models of the complicated and unknown processes of the noise generations and mixings. This technique facilitates understanding of the precise behaviors and properties of target nanoparticles by using the multistage narrowing nanochannel measurement.

2   Related Work and Research Objective

Frequency filtering [2], Fourier and wavelet transforms [3], and Kalman filtering [4] have been frequently used to date for signal noise reduction. They are effective for cases in which noise generation and mixing processes have linearity or when they can be well approximated using simple known models. Nevertheless, these noise reduction approaches are not applied effectively to complicated and unknown noise processes such as many advanced measurement problems including the multistage narrowing nanochannels.

To address this difficulty, research into approaches designated as Noise2Clean have been developed where known objects are preliminarily measured, then the relation between the object with its correct answer and its noisy measurement result is learned by deep learning. Eventually, the correct objects are estimated inversely from the measurement results by using the learned deep neural network model as a noise filter. This technique has been broadly applied to many problems including reduction of regular noise, JPEG compression noise and text-missing noise [5] super-resolution image estimation [6], image color estimation [7], and image restoration [8]. However, in many advanced measurement problems, target objects with correct answers are hardly obtained, thus the cases to which this Noise2Clean approach is applicable are very limited. In our problem to measure the electric current waveform for nanoparticles passing using a multistage narrowing nanochannel, the correct waveform cannot be directly ascertained in principle.

On the other hand, a recently proposed unsupervised approach designated as Noise2Noise [9] obtains a noise reduction model using U-net, if multiple independent measurements with noise are available on same objects, even for cases in which a correct answer for each object is unknown. A U-net model is a fully convolutional AutoEncoder specially designed to retain fine structures of the input signal in the output. In many advanced measurement problems, acquiring data for multiple independent measurements of the same objects or the same type of objects is not difficult. Therefore, this approach might pave the way to construction of a highly accurate noise reduction model in advanced measurements including complex noise generation and mixing processes.

Based on these technical advancement, we set our target objective in this study to assess two types of Noise2Noise models namely U-net models and standard Convolutional AutoEncoder (CAE) models to reconstruct the electric current waveforms of particle passing in a nanochannel from their highly noisy measurements. We show that a CAE model reproduces more effectively denoised waveform without loosing the waveform structure than the U-net models presented in the original Noise2Noise study [9].

3   Outline of Multistage Narrowing Nanochannel

A nanochannel device comprises a groove designated as a nanochannel dug into a flat silicon plate, an electrolyte solution to fill this groove, plus and minus electrodes provided at the inlet and outlet of the groove. When an electric voltage is applied across both electrodes, an ion current flows through the nanochannel. When a nanoparticle of the measurement object is introduced, the ion current flow passing through the nanochannel is prevented by temporary and partial blocking of the groove. Passing of nanoparticles is visible as a current waveform by one particle unit if the time change of this ion current is recorded.

As presented in Fig. 1, in the multistage narrowing nanochannel device targeted in this research, a five stage narrowing part is provided along with the flow path of nanoparticles. Changes in the migration velocity of the particles can be estimated by closely analyzing the current waveform if the waveform is accurately measured without noise. For example, consider the case in which the target particle is a biological cell. The cell growth control mechanism can be elucidated based on the change of the individual cell weight estimated from the migration velocity measured by this device in conjunction with theoretical calculations of electrophoresis.

Figure 2(a) presents the expected measurement current waveform for the case in which one particle passes a five-stage nanochannel. When a nanoparticle passes through the nanochannel, the whole ion current waveform is depressed, since the flow path of the ion current is kept partially closed during the passage. In addition, it is expected that five small depressions are apparent at the bottom of the entire depression, because the closure of the nanochannel is intensified at the instant when a nanoparticle passes a narrowing part of the nanochannel.

On the other hand, Fig. 2(b) presents an example of an actually measured electric current waveform. The current waveform measured by a multistage narrowing nanochannel device includes much noise introduced through its complex and unknown generation and mixing processes. Ascertaining the correct

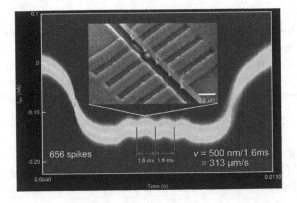

Fig. 1. A multistage narrowing nanochannel and an electric current waveform.

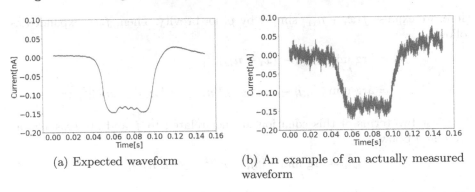

(a) Expected waveform

(b) An example of an actually measured waveform

Fig. 2. Output current waveforms of a multistage narrowing nanochannel.

migration of nanoparticles by using this measurement result is difficult, since the detailed profiles of the waveform are hardly preserved. Accordingly, effective noise reduction for the measured current waveforms is indispensable for the practical use of the multistage narrowing nanochannel devices.

4 Principle and Implementation of Noise2Noise Deep Learning

4.1 Principle

In the framework of Noise2Clean, pairs of correct feature vector x_i and its feature vector $x_i + n_i$ distorted by noise n_i are presumed to be given for many instances i. In this case, noise reduction model f_n is obtained by the following formula:

$$f_n = \arg \min_f \mathbb{E}[||f(x_i + n_i) - x_i||^2] \tag{1}$$

in supervised learning.

On the other hand, Noise2Noise [9] is a method to obtain a noise reduction model of the target instances in unsupervised fashion. Every instance i is assumed to be measured at least more than or equal to twice. Then, a pair of feature vectors $\boldsymbol{x}_i + \boldsymbol{n}_{i1}$ and $\boldsymbol{x}_i + \boldsymbol{n}_{i2}$ measured with independent noise $\boldsymbol{n}_{i1}$ and $\boldsymbol{n}_{i2}$ is given for every instance i. A noise reduction model $\tilde{f}_n$ is obtained from these measurement pairs without knowing the correct $\boldsymbol{x}_i$ as follows.

$$\tilde{f}_n = \arg\min_f \mathbb{E}_i[||f(\boldsymbol{x}_i + \boldsymbol{n}_{i1}) - (\boldsymbol{x}_i + \boldsymbol{n}_{i2})||^2]. \tag{2}$$

In the rest of this subsection, we show that $\tilde{f}_n$ given by Eq. (2) agrees with f_n given by Eq. (1) under the assumptions that the noise of every measurement is independent from the others and the expected noise average is zero, *i.e.*,

$$\mathbb{E}_i[\boldsymbol{n}_i] = \boldsymbol{0}. \tag{3}$$

Here, we express $f(\boldsymbol{x}_i + \boldsymbol{n}_{i1})$ simply by $\boldsymbol{y}_i$ for brevity. Then, $\tilde{f}_n$ is expanded as follows.

$$\tilde{f}_n = \arg\min_f \mathbb{E}_i[||\boldsymbol{y}_i - (\boldsymbol{x}_i + \boldsymbol{n}_{i2})||^2]$$

$$= \arg\min_f \mathbb{E}_i[||\boldsymbol{y}_i - \boldsymbol{x}_i||^2 - \boldsymbol{y}_i^T \boldsymbol{n}_{i2} - \boldsymbol{x}_i^T \boldsymbol{n}_{i2} + ||\boldsymbol{n}_{i2}||^2].$$

The last two terms of this equation are not related to f and are erasable, as presented below.

$$\tilde{f}_n = \arg\min_f [\mathbb{E}_i ||\boldsymbol{y}_i - \boldsymbol{x}_i||^2 - \mathbb{E}_i(\boldsymbol{y}_i^T \boldsymbol{n}_{i2})]. \tag{4}$$

Furthermore, as noise $\boldsymbol{n}_{i2}$ is independent of feature vector $\boldsymbol{x}_i$, noise $\boldsymbol{n}_{i1}$, and form f, $\boldsymbol{n}_{i2}$ is also independent of $\boldsymbol{y}_i = f(\boldsymbol{x}_i + \boldsymbol{n}_{i1})$. Therefore, simultaneous distribution of $\boldsymbol{y}_i$ and $\boldsymbol{n}_{i2}$ agrees with

$$P(\boldsymbol{y}_i, \boldsymbol{n}_{i2}) = P(\boldsymbol{y}_i)P(\boldsymbol{n}_{i2}).$$

By using this relation and Eq. (3),

$$\mathbb{E}_i(\boldsymbol{y}_i^T \boldsymbol{n}_{i2}) = \iint \boldsymbol{y}_i^T \boldsymbol{n}_{i2} \, P(\boldsymbol{y}_i, \boldsymbol{n}_{i2}) \, d\boldsymbol{y}_i d\boldsymbol{n}_{i2}$$

$$= \int \boldsymbol{y}_i^T P(\boldsymbol{y}_i) d\boldsymbol{y}_i \int \boldsymbol{n}_{i2} \, P(\boldsymbol{n}_{i2}) d\boldsymbol{n}_{i2}$$

$$= \mathbb{E}_i(\boldsymbol{n}_i) \int \boldsymbol{y}_i^T P(\boldsymbol{y}_i) d\boldsymbol{y}_i = 0$$

is obtained. Then, the second term of Eq. (4) is dropped off, and

$$\tilde{f}_n = \arg\min_f \mathbb{E}_i[||f(\boldsymbol{x}_i + \boldsymbol{n}_{i1}) - \boldsymbol{x}_i||^2]$$

is obtained, which agrees with Eq. (1).

Accordingly, even if correct measurements of the targets without noise is not obtained, one can get a noise reduction model equivalent to a supervised case using the Noise2Noise method with noisy measurements of the targets only.

4.2 Implementation

To obtain an appropriate noise reduction model by the principle of Noise2Noise for measurement results where many noises are added through complicated and unknown generation and mixing processes, one must use a model having high learning capability. Accordingly, we employ deep learning in this study. Regarding the network model to be used, one that produces noise reduction output y_i of time series waveform that is the same as input time series waveform x_i is suitable. In this study, U-net models used by Lehtinen et al. [9] and Convolutional AutoEncoder (CAE) models are applied.

CAE used here has 15,000 dimensional input equivalent to the dimension of time series waveform vector x_i. The encoder has 4 layers, and each layer consists of multiple convolution channels. The input x_i is encoded to the elements ranged in 625 dimensions $\times 64$ channels through this encoder. These elements are further decoded into a denoised 15,000 dimensional time series waveform vector y_i via symmetrical four-layer deconvolution. While the dimension reduction in the CAE is conducted by using pooling in many cases, it is highly probable that a part of features of the input information is lost by the application of pooling to one-dimensional time series input, such as measured current waveform. Therefore, the dimension reduction is performed by strides in the convolution without using pooling in our approach. Particularly, we let the kernel width of each layer be 10 and the stride width be 2–3. For the activation function of each convolution layer and deconvolution layer, the ReLu function is used, except for the final layer of the decoding part. The Leaky-ReLu function, by which learning by back propagation is advanced easily, is used for the final layer of the decoding part.

U-net [9], which Lehtinen et al. used as a noise reduction model for Noise2Noise, has a Skip Connection between each convolution layer output of the encoder part and its corresponding deconvolution layer input of decoder part. These connections highlight the input position information in the time series waveform and improve convergence of learning. In our study, a U-net is designed by adding the Skip Connection to bridge every encoder layer and its corresponding decoder layer in the aforementioned CAE.

5 Training of Noise Reduction Model

5.1 Training Data, Training Procedures, and Performance Verification

Figure 3 presents the training process of a noise reduction model for the multistage narrowing nanochannel and its verification procedure. 500 current waveforms obtained by independently measuring nearly the same particle passing phenomena in a five-stage narrowing nanochannel are used for the training shown in the upper picture. Each current waveform is a time series vector comprising 15,000 steps measured for 0.015 sec by 1 MHz sampling. Each waveform vector is standardized so that the waveform vector average is 0 and its variance is 1.

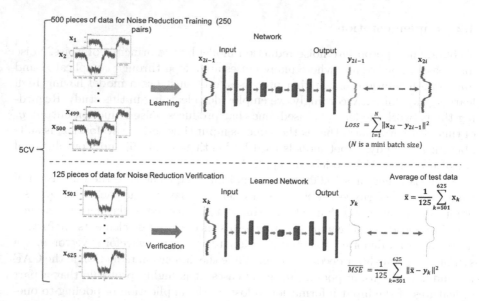

Fig. 3. Training and verification of noise reduction model using waveform data.

These 500 vectors are subjected to pairing at random, and 250 pairs of the independently measured waveform vector $\{(\boldsymbol{x}_{2i-1}, \boldsymbol{x}_{2i})\}(i = 1, \cdots, 250)$ are used for the training.

The mean square error calculated from the 250 pairs of the waveform vectors is used as a total loss function for both CAE and U-net trainings. Adam algorithm [10] with mini batch size 16 is used as the optimizer in each network training. Early Stopping at epoch number 50 is applied to avoid overtraining.

The lower part of Fig. 3 shows the performance verification procedure. 125 standardized waveform vectors $\boldsymbol{x}_k (k = 501, \cdots, 625)$ which are not used for the above training are input to each trained model to obtain denoised current waveform $\boldsymbol{y}_k$. Because the correct target waveform presented in Fig. 2(a) is unknown, we employ the total average waveform $\overline{\boldsymbol{x}}$ of these 125 $\boldsymbol{x}_k (k = 501, \cdots, 625)$ as a surrogate target waveform. We evaluate the noise reduction performance by using the mean square error (MSE) of 125 pieces of $\boldsymbol{y}_k$ and $\overline{\boldsymbol{x}}$. This verification procedure is repeated in 5-hold cross validation where the 500 waveform vectors for the training and the 125 waveform vectors for the verification are interchanged sequentially. We use the total average of the mean square errors (MSE) over the 5-hold cross validation as the performance index of a noise reduction model.

5.2 Methods for Performance Comparison

As performance comparison targets, this study uses noise reduction methods using Fourier transform and wavelet transform, which are frequently used for signal processing. Here, the Fourier transform is used to assess the noise removal from a waveform in a frequency domain, while the wavelet transform is used to assess the noise removal using flexible local windows of both time and frequency domains.

Figure 2(b) shows that main noise components of the five stage narrowing nanochannel have high-frequencies. Therefore, in the Fourier transform based noise reduction, the measured electric current waveform is converted to the frequency domain by Fast Fourier transform (FFT), and the high-frequency component above a frequency threshold is cut. Then the noise-reduced electric current waveform is obtained by its inverse FFT. The best cutoff frequency threshold is found using a line search to minimize the mean square error (MSE) between the filtered waveforms of the training vectors $x_i (i = 1, \cdots, 500)$ and the average waveform $\overline{x}$ of $x_i (i = 1, \cdots, 500)$.

In the wavelet transform based noise reduction, assuming that the primary components of the correct electric current waveform are consolidated to the wavelet component having greater strength, small strength wavelet components of less than a certain threshold are excluded as noise components. Wavelet transform $w_\psi(a, b)$ of the signal waveform $f(t)$ is defined as

$$w_\psi(a, b) = \int_{-\infty}^{\infty} \frac{1}{\sqrt{a}} f(t) \psi \left(\frac{t - b}{a} \right) dt.$$

where a and b are scale and shift parameters, respectively. $\psi(\cdot)$ is the wavelet base function. The original signal waveform $f(t)$ is obtained via inverse wavelet transform.

$$f(t) = \frac{1}{C_\psi} \int_0^{\infty} \int_{-\infty}^{\infty} \frac{1}{a^2} w_\psi(a, b) \psi_{a,b}(t) da db, \tag{5}$$

where C_ψ is a constant determined by the type of a wavelet base function. Therein, $|w_\psi(a, b)|$ shows the wavelet component strength of the parameters a and b. Therefore, $|w_\psi(a, b)|$ of strength less than a certain threshold is enforced to be 0 for the noise reduction. Then a inverse wavelet transform is applied and a denoised electric current waveform is obtained. Regarding the wavelet base functions, Haar, Daubechies, Symlets, Coiflets, Biorthogonal, Reverse biorthogonal, Discrete Meyer, Mexican hat, Morlet, and Gaussian are tested. With a similar manner to that used for Fourier transformation, grid search is performed to select the best combination of a cut off threshold of the strength and a wavelet base function by minimizing the mean square error (MSE) between the filtered waveforms of the training vectors $x_i (i = 1, \cdots, 500)$ and the average waveform $\overline{x}$ of $x_i (i = 1, \cdots, 500)$.

5.3 Verification Results

Numerical experiments were performed using a computer equipped with a CPU (8 core processor with Xeon W-2145 3.7 GHz; Intel Corp.) and a GPU (TITAN RTX x 2NVLink; NVIDIA). The program development environment was Python 3 on Linux. The pytorch learning library was used.

The MSE in Ampere unit of electric current obtained by each method is presented in Table 1. The optimum threshold of the frequency of Fourier transform 534 Hz. The optimum wavelet base function was Daubechies. Compared with the conventional method by Fourier transform and wavelet transform, the Noise2Noise models using both CAE and U-net reconstruct the waveform more similar to the surrogate target waveform $\bar{x}$.

An example of a raw measured waveform, which has been used for the verification, is shown in Fig. 4 (a). The red lines in Fig. 4 (b)–(e) show the surrogate target waveforms, which are the average waveforms over all 125 electric current waveforms used in the verification. The blue lines n Fig. 4 (b)–(e) present the waveforms denoised by the denoising methods, respectively. Fine variation of the electric current when one particle passes through five-stage narrowing parts, as depicted by the red lines in Fig. 4 (b)–(e), cannot be extracted by Fourier transform (b) and wavelet transform (c), whereas CAE (d) and U-net (e) can do clearly.

In terms of mean square error (MSE), the U-net slightly outperformed the CAE in this verification study as indicated in Table 1. However, the U-net retained some noise component in its waveform reconstruction as shown by the light blue line in Fig. 4 (f), while the CAE provided an almost completely denoised waveform as shown by the blue line in Fig. 4 (f). We observed the same behaviors of the U-net and the CAE in the other waveforms. In this regard, the CAE is more suitable for the effective noise reduction without loosing the waveform structure.

Table 1. Comparison of Mean Square Error (MSE).

Model	MSE
Fourier transform	1.5285×10^{-22}
Wavelet transform	1.2036×10^{-22}
CAE	1.0456×10^{-22}
U-net	1.0395×10^{-22}

6   Verification for a Two-Particle Passing Waveform

In cases that multiple nanoparticles sequentially pass through the five stage narrowing part of the nanochannel at the same time, the electric current waveform measured by the nanochannel device show a complex pattern reflecting the time differences of these particle movements and the interactions between these particles. Figure 5(a) presents one example obtained when two nanoparticles flow simultaneously through the five stage narrowing nanochannel presented in Fig. 1. When the first nanoparticle enters the nanochannel, the ion current is reduced by partial closure of the channel. When the second nanoparticle enters, the ion

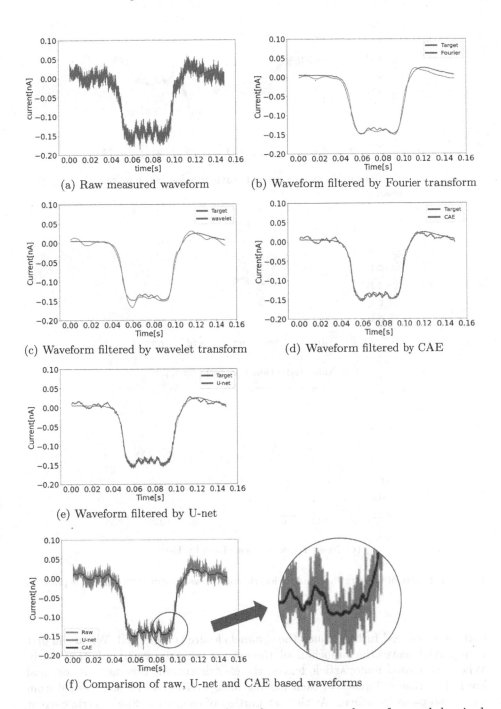

(a) Raw measured waveform

(b) Waveform filtered by Fourier transform

(c) Waveform filtered by wavelet transform

(d) Waveform filtered by CAE

(e) Waveform filtered by U-net

(f) Comparison of raw, U-net and CAE based waveforms

Fig. 4. Comparison of a raw measured waveform, an averaged waveform and denoised waveforms.

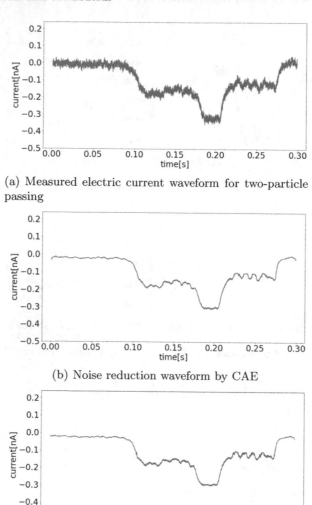

(a) Measured electric current waveform for two-particle passing

(b) Noise reduction waveform by CAE

(c) Noise reduction waveform by U-net

Fig. 5. Example of noise reduction for electric current waveform at two-particle passing.

current is reduced further, since the channel closure is increased. When the first nanoparticle leaves the nanochannel, the ion current is recovered only that much. When the second nanoparticle leaves, the electric current reverts to its original level. The time difference of each passing nanoparticle can be ascertained from such a stepwise waveform. At the flat portion of each step, fine electric current change, corresponding to each nanoparticle's passage through a narrowing part, is hidden in the noise.

One can ascertain the precise behaviors of these particles if the accurate electric current waveform can be estimated from measured electric waveforms that include much noise. For this purpose, we applied both CAE and U-net based noise reduction models obtained in the former section to the measured electric current waveform presented in Fig. 5(a). Figure 5(b) presents results obtained from the CAE noise reduction model, and (c) presents results obtained by the U-net noise reduction model. Both models were able to reduce the noise of the waveform effectively and derived fine variations of electric current that occurs when each particle passes through the narrowing part. However, the U-net result slightly more retains the noise in its output than the CAE result as argued in the former section.

7   Conclusion

In this study, we applied the Noise2Noise method to effectively reduce noise from the measured electric current waveforms in which noise was generated and mixed by complex and unknown processes in a multistage narrowing nanochannel. Noise reduction models using CAE and U-net were trained in an unsupervised manner without knowing the correct electric current waveform measuring one-particle passing. These noise reduction models were confirmed as being capable of noise reduction without losing details of electric current waveform information as compared with conventional signal processing methods using Fourier transform and wavelet transform. It has also been confirmed that effective noise reduction can be achieved from a measured electric current waveform when multiple particles pass through the nanochannel simultaneously. In the application of the Noise2Noise framework to the multistage narrowing nanochanne, we observed the superior denoising performance of the CAE rather than the U-net. The CAE based noise reduction model could denoise the current waveform almost completely.

Further applications of the Noise2Noise method to other advanced measurement problems will be attempted in our future work, and we will further seek new principles for measurement information processing incorporating the Noise2Noise method.

Acknowledgement. This research is supported by JST CREST JMPJCR1666.

References

1. Tsutsui, M., et al.: High-throughput single nanoparticle detection using a feed-through channel-integrated nanopore. Nanoscale **43**, 20475–20484 (2019)
2. Basty, A.: Shenoy: Introduction to Digital Signal Processing and Filter Design. Wiley, Hoboken (2005)
3. Ryan, O.: Linear Algebra, Signal Processing, and Wavelets - A Unified Approach. Springer, Heidelberg (2019). https://doi.org/10.1007/978-3-030-02940-1
4. Grewal, M.S., Andrews, A.P.: Kalman Filtering: Theory and Practice with MATLAB. Wiley - IEEE (2014)

5. Mao, X.-J., Shen, C., Yang, Y.-B.: Image restoration using convolutional auto-encoders with symmetric skip connections. In: Proceedings of Thirtieth Conference on Neural Information Processing Systems, NIPS 2016 (2016)
6. Ledig, C., et al.: Photo-realistic single image super-resolution using a generative adversarial net-work. In: Proceedings of Computer Vision and Pattern Recognition, CVPR 2017, pp. 105–114 (2017)
7. Zhang, R., Isola, P., Efros, A.A.: Colorful image colorization. In: Leibe, B., Matas, J., Sebe, N., Welling, M. (eds.) ECCV 2016. LNCS, vol. 9907, pp. 649–666. Springer, Cham (2016). https://doi.org/10.1007/978-3-319-46487-9_40
8. Iizuka, S., Simo-Serra, E., Ishikawa, H.: Globally and locally consistent image completion. ACM Trans. Graph. **36**(4), 107:1–107:14 (2017)
9. Lehtinen, J., et al.: Noise2Noise: learning image restoration without clean data. In: Proceedings of the 35th International Conference on Machine Learning, ICML 2018, PMLR, vol. 80, pp. 2965–2974 (2018)
10. Kingma, D.P., Ba, J.: Adam: a method for stochastic optimization. In: Proceedings of the 3rd International Conference on Learning Representations, ICLR 2015 (2015)

Classification Bandits: Classification Using Expected Rewards as Imperfect Discriminators

Koji Tabata[1,3]([✉]), Atsuyoshi Nakumura[2], and Tamiki Komatsuzaki[1,3]

[1] Institute for Chemical Reaction Design and Discovery, Hokkaido University, Sapporo, Japan
ktabata@es.hokudai.ac.jp
[2] Graduate School/Faculty of Information Science and Technology, Hokkaido University, Sapporo, Japan
[3] Research Center of Mathematics for Social Creativity Research Institute for Electronic Science, Hokkaido University, Sapporo, Japan

Abstract. A classification bandits problem is a new class of multi-armed bandits problems in which an agent must classify a given set of arms into positive or negative depending on whether the number of bad arms are at least N_2 or at most $N_1(< N_2)$ by drawing as fewer arms as possible. In our problem setting, bad arms are imperfectly characterized as the arms with above-threshold expected rewards (losses). We develop a method of reducing classification bandits to simpler one threshold classification bandits and propose an algorithm for the problem that classifies a given set of arms correctly with a specified confidence. Our numerical experiments demonstrate effectiveness of our proposed method.

Keywords: Multi-armed bandits · Threshold bandits · Thompson sampling

1 Introduction

How to determine the presence or extent of a disease from biopsy under the existence of some uncertainty is of crucial importance in life science. Let us consider the following cancer diagnosis problem in which a doctor has to diagnose whether a certain patient has cancer or not from given his/her K cells: if the number of cancer cells N is negligible ($N \leq N_1$), then the doctor can diagnose that the patient does not have cancer, but if it is non-negligible ($N \geq N_2 > N_1$), the doctor should diagnose that the patient does. One of the cancer cell diagnosis methods is the classification of cells in terms of a set of Raman spectra[1] averaged over each cell [8]. However, Raman measurements require more than ten hours

---

[1] Histopathologists usually diagnose whether cells are of cancer or not by inspecting their morphological characteristics with a human bias, but Raman measurements are considered to enable more reliably to judge the cell states.

© Springer Nature Switzerland AG 2021
M. Gupta and G. Ramakrishnan (Eds.): PAKDD 2021 Workshops, LNAI 12705, pp. 57–69, 2021.
https://doi.org/10.1007/978-3-030-75015-2_6

for one hundred cells, by scanning point illumination to single cells along which Raman spectra are acquired in time to time. Thus, interactive measurement depending on Raman spectra obtained so far is a key to realize fast cell diagnosis. Such interactive measurement can be formulated as a bandit problem treated in this paper by regarding each cell as an arm, and letting Raman spectra sampling from each cell correspond to an arm draw.

In this bandit problem, doctor cannot always conclude whether the number of cancer cells is negligible or not correctly due to two different types of uncertainty. The first type of uncertainty is the variance of reward (cancer index calculated from sampled Raman spectra) obtained by each arm draw, which has been extensively studied in the area of statistics. The second type of uncertainty is the imprecision (imperfect positive predictive value) of the true expected reward of each arm (cancer index averaged over each cell) and this frequently happens in real situation. In fact, Raman spectra averaged over each cell was reported to be classified into cancer or normal cells with about 85% accuracy [8]. While we can obtain as much accurate value as we want by taking enough number of samples for the first type of uncertainty, the second type of uncertainty cannot be reduced. To the best of our knowledge, this paper is the first one taking account of the second type of uncertainty in the context of bandits.

In this paper, we study a pure exploration K-armed bandit problem, named *classification bandit problem*, in which an agent must classify a given set of K arms into "negative" or "positive" depending on whether the number of bad arms is at most N_1 or at least $N_2(> N_1)$, respectively, within given allowable failure probabilities δ_N and δ_P by drawing as small number of arms as possible. In our formulation, the mean reward μ_i of each arm i is assumed to be an imperfect discriminator of badness; $\mu_i \geq \theta$ holds for each bad arm i with probability $p_{TP} > 0.5$ and $\mu_i < \theta$ holds for each good arm i with probability $p_{TN} > 0.5$.

We show that the classification bandit problem with parameters $p_{TP}, p_{TN}, \delta_N,$ δ_P, N_1, N_2 can be reduced to the problem, named *one-threshold classification bandit problem*, which has only one threshold λ instead of two thresholds N_1 and N_2 and one allowable failure probability δ instead of δ_P and δ_N and is free from the second type of uncertainty. Our reduction is not always possible, and we show the condition of the reduction and how to calculate λ and δ from $p_{TP}, p_{TN}, \delta_N, \delta_P, N_1, N_2$. For the one-threshold classification bandit problem, we propose an algorithm and prove its correctness for any arm selection policy. We also propose a Thompson-sampling-based arm selection policy for this algorithm, which demonstrates a faster stopping time of the algorithm than UCB-based and Successive-Elimination-based arm selection policies.

Related Works

One-threshold classification bandits problem is regarded as a kind of pure exploration bandit problem. Complexity analysis of this type of problem is performed by minimizing error probability under a fixed budget or minimizing the number of samples under a fixed confidence. In this paper we focus on the fixed confidence setting. The most studied pure exploration bandit problem is best arm identification whose objective is to find the k highest expected reward arms for a given k. For the best arm identification problem, Audibert et al. [1] proposed an algorithm based on successive elimination that eliminates the worst arm one by one from candidates of the best arm. Later, more efficient algorithms that are not based on elimination such as LUCB [3] and UGapE [2] were proposed.

Instead of the highest expected reward arms, Locatelli et al. [6] proposed an algorithm for a thresholding bandit problem in which an agent has to output all the arms with the expected reward higher than a given threshold. Kano et al. [4] formulated the good arm identification problem whose task is to find λ arms whose expected rewards are above a given threshold, for a given λ, if at least such λ arms exist, or to find all such arms otherwise. Kaufmann et al. [5] and Tabata et al. [10] independently studied a problem to decide whether at least one arm exists or not, whose expected reward is above the threshold, in which precise identification of such arms is not necessarily required.

A question of 'how one can derive accurate decision under the condition that only qualitative test results would be obtained' is one of the most intriguing subjects in the area of fault detection of systems. In many cases, several kinds of tests are assumed to be given explicitly with their false-positive, and the false-negative rates. Nachla et al. [7] treated a problem to design the permutation of tests in order to decrease the total cost of, e.g., quality inspections, repairs of good components in products, and dispositions of no-fault systems, and proposed heuristics to solve the problem. Raghavan et al. [9] proposed a method to decide an optimal test sequence with qualitative tests by using dynamic programming.

To our best knowledge, there exists no research that deals with the problem of accurate decision in terms of qualitative test results in the context of bandits algorithm. In this paper, we present an algorithm to transform classification bandits based on qualitative tests with nonnegligible false-positive and false-negative rates into one-threshold bandits problem which can address the transformability to one threshold bandits and design the threshold and the error rate to meet the given allowed error rates.

2 Problem Formulation

We study a variant of a K-armed bandit problem defined as follows. An agent is given a set of K arms that is composed of *bad* and *good* arms. No information about which arm is bad or good is directly provided to the agent. However, the agent can get values $X_i(1), X_i(2), \cdots$ of an indicator to represent badness for each arm i by drawing it repeatedly, where $X_i(n)$ denotes the value obtained by the nth draw of arm i. For each arm i, we assume that $X_i(1), X_i(2), \cdots$ are i.i.d. random variables whose distribution is denoted as ν_i with mean μ_i (whose distribution and value cannot be known *a priori*).

For a given threshold θ, an arm i satisfying $\mu_i \geq \theta$ ($\mu_i < \theta$) is defined as *positive (negative) arm*. We consider the case that arm's expected indicator value μ_i is an imperfect discriminator of its badness or goodness; a bad arm is positive with probability p_{TP} and a good arm is negative with probability p_{TN}. We also assume that, whether each arm is positive or negative, is independent of any different arms i and j, e.g., $P[\text{arms } i \text{ and } j \text{ are positive}] = P[\text{arm } i \text{ is positive}]P[\text{arm } j \text{ is positive}]$.

At every time step $t = 1, 2, \ldots$, an agent chooses one of the K arms $i_t \in \{1, 2, \ldots, K\}$ and gets an indicator value $X_{i_t}(n_{i_t}(t))$ of badness that is drawn from distribution ν_{i_t}, where $n_i(t)$ is the number of times arm i has been drawn by time step t. The agent's task is to conclude whether the given set of K arms contains non-negligible number of bad arms (i.e., a number of bad arms enough to judge "positive") or not by drawing arms as few times as possible. We can formulate the problem as follows.

Problem 1 (classification bandits). For given $p_{\mathrm{TP}}, p_{\mathrm{TN}} \in (0.5, 1]$ and $\delta_P, \delta_N \in (0, 0.5)$, output "negative" with probability at least $1 - \delta_N$ if the number of bad arms is less than or equal to N_1, and output "positive" with probability at least $1 - \delta_P$ if the number of bad arms is larger than or equal to N_2 by drawing as small number of arms as possible.

Note that there is no requirement for the output when the number of bad arms is larger than N_1 and lower than N_2.

3 Problem Reduction

3.1 Reduction Theorem

In this section, we show how to reduce classification bandits, which contain uncertainty derived from probabilities p_{TP} and p_{TN}, to the following one-threshold classification bandits, which is free from such uncertainty.

Problem 2 (one-threshold classification bandits). For a given $\lambda \in \mathbb{N}$ and $\delta > 0$, output "negative" with probability at least $1 - \delta$ if the number of positive arms is less than λ, and output "positive" with probability at least $1 - \delta$ if the number of positive arms is at least λ by drawing as small number of arms as possible.

In this problem setting, the number to be identified is not the number of bad arms but that of positive arms, that is, we only consider the uncertainty due to reward variance to solve this problem. We will introduce the algorithm to solve this reduced problem in the next section. Here we explain how such reduction is possible.

Given the number of arms K, the number of bad arms $N \in [0, K]$, probability $p_{\mathrm{TP}} > 0.5$ with which a bad arm is positive, and probability $p_{\mathrm{TN}} > 0.5$ with which a good arm is negative, the probability-generating function $G^{(N,K)}(t)$ for the number of positive arms is expressed as

$$G^{(N,K)}(t) = (p_{\mathrm{TP}}t + (1 - p_{\mathrm{TP}}))^N ((1 - p_{\mathrm{TN}})t + p_{\mathrm{TN}})^{K-N},$$

because of the arm's independency.

Let $c_d^{(N,K)}$ denote the coefficient of t^d in $G^{(N,K)}(t)$. Then, $c_d^{(N,K)}$ is the probability that the number of positive arms is d in the case that the given set of K arms contains just N bad arms.

The following proposition is used to prove Lemma 1.

Proposition 1. $\sum_{j=0}^{\ell-1} c_j^{(N,K)}$ is a weakly decreasing function on N and $\sum_{j=\ell}^{K} c_j^{(N,K)}$ is a weakly increasing function on N.

Proof. Omitted due to space limitations.   □

Let X denote the number of positive arms. Then, $P[X \geq \ell | N = i]$ and $P[X < \ell | N = i]$, probabilities of $X \geq \ell$ and $X < \ell$ under the condition of $N = i$, are $\sum_{j=\ell}^{K} c_j^{(i,K)}$ and $\sum_{j=0}^{\ell-1} c_j^{(i,K)}$, respectively.

From Proposition 1, we can have the following lemma.

Lemma 1. *Let X and N be the number of positive arms and the number of bad arms, respectively. Then, the following two inequalities hold:*

$$P[X \geq \ell | N \leq N_1] \leq P[X \geq \ell | N = N_1] \text{ and}$$
$$P[X < \ell | N \geq N_2] \leq P[X < \ell | N = N_2].$$

Proof. By Proposition 1, for $i \leq N_1$,

$$P[X \geq \ell | N = i] = \sum_{j=\ell}^{K} c_j^{(i,K)} \leq \sum_{j=\ell}^{K} c_j^{(N_1,K)} = P[X \geq \ell | N = N_1]$$

holds, and thus

$$P[X \geq \ell | N \leq N_1] = \frac{\sum_{i=0}^{N_1} P[X \geq \ell | N = i] P[N = i]}{\sum_{i=0}^{N_1} P[N = i]}$$

$$\leq \frac{\sum_{i=0}^{N_1} P[X \geq \ell | N = N_1] P[N = i]}{\sum_{i=0}^{N_1} P[N = i]} = P[X \geq \ell | N = N_1].$$

We can show the second inequality similarly.   □

This implies that, e.g., the failure probability $P[X \geq \ell | N \leq N_1]$, that is, under the condition of $N \leq N_1$ to be identified as negative, the classifier answers "positive $(X \geq \ell)$", can be upperbounded by $P[X \geq \ell | N = N_1]$. By the above lemma, the following reduction theorem can be proved.

Theorem 1 (reduction theorem). *For classification bandit problem with $p_{TN}, p_{TP} \in (0.5, 1]$, $\delta_P, \delta_N \in (0, 0.5)$, $0 \leq N_1 < N_2 \leq K$, consider the one-threshold classification bandit problem with*

$$\lambda = \arg \max_{\ell} \delta(\ell, p_{TN}, p_{TP}, N_1, N_2, K), \tag{1}$$

$$\delta = \delta(\lambda, p_{TN}, p_{TP}, N_1, N_2, K), \tag{2}$$

where $\delta(\ell, p_{TN}, p_{TP}, N_1, N_2, K) = \min\left(\delta_N - \sum_{j=\ell}^{K} c_j^{(N_1,K)}, \delta_P - \sum_{j=0}^{\ell-1} c_j^{(N_2,K)}\right).$

In the case with $\delta > 0$, classification bandit problem associated with nonzero false-positive and false-negative rates can be reduced to a one-threshold bandit problem.

Proof. Assume that $\delta > 0$ and algorithm A is an algorithm for one-threshold bandit problem with λ and δ defined Eqs. (1) and (2). Consider the case that the number of bad arms N is at most N_1, which is the case that the algorithm is desired to output "negative." Let X be the number of positive arms. The failure probability that algorithm A outputs "positive" falsely is upper-bounded by $P[X \geq \lambda | N \leq N_1] + \delta$. By Eq. (2) and Lemma 1, $\delta \leq \delta_N - P[X \geq \lambda | N = N_1] \leq \delta_N - P[X \geq \lambda | N \leq N_1]$ holds. Thus, $P[X \geq \lambda | N \leq N_1] + \delta \leq \delta_N$ holds. For the case that the number of bad arms N is at least N_2, algorithm A can be proved similarly to output "negative" with probability at most δ_P.   □

3.2   Reducible Parameter Region

In the one-threshold classification bandits problem that is reduced from classification bandits problem, the value of confidence parameter δ calculated by reduction theorem, that is, $\delta = \max_\ell \min\left(\delta_N - \sum_{j=\ell}^{K} c_j^{(N_1,K)}, \delta_P - \sum_{j=0}^{\ell-1} c_j^{(N_2,K)}\right)$, significantly affects sample complexity of the problem, and the problem is not solvable if this δ is non-positive. In the followings, we derive an approximate boundary $\delta = 0$ between solvable and unsolvable (N_1, N_2)-region for fixed $p_{TP}, p_{TN}, \delta_P, \delta_N$.

Let $X(m)$ be the number of positive arms when the number of bad arms is m. Then, $X(m)$ can be expressed as $X(m) = X_B + X_G$ using $X_B \sim B(m, p_{TP})$ (B:binomial distribution) and $X_G \sim B(K-m, 1-p_{TN})$. By law of large numbers, $B(m, p_{TP}) \approx N(m p_{TP}, m p_{TP}(1-p_{TP}))$ and $B(K-m, 1-p_{TN}) \approx N((K-m)(1-p_{TN}), (K-m)p_{TN}(1-p_{TN}))$ when $m p_{TP}, m(1-p_{TP}), (K-m)(1-p_{TN})$ and $(K-m)p_{TN}$ are enough large (e.g. ≥ 5). By reproductive property of normal distribution, $X(m) \sim N(\mu_m, \sigma_m^2)$ holds for $\mu_m = m p_{TP} + (K-m)(1-p_{TN})$ and $\sigma_m^2 = m p_{TP}(1-p_{TP}) + (K-m)p_{TN}(1-p_{TN}))$.

By the Polya's approximation $\frac{1}{\sqrt{2\pi}} \int_0^x \exp\left(-\frac{t^2}{2}\right) dt \approx \frac{1}{2}\sqrt{1 - \exp\left(-\frac{2}{\pi}x^2\right)}$, $\mathbb{P}[X(m) \geq \mu_m + \alpha_\delta \sigma_m] \approx \delta$ and $\mathbb{P}[X(m) \leq \mu_m - \alpha_\delta \sigma_m] \approx \delta$ hold for $\alpha_\delta = \sqrt{\frac{\pi}{2} \ln \frac{1}{1-(1-2\delta)^2}}$. Thus, $\mathbb{P}[X(N_1) \geq \mu_{N_1} + \alpha_{\delta_N} \sigma_{N_1}] \approx \delta_N$ and $\mathbb{P}[X(N_2) \leq \mu_{N_2} - \alpha_{\delta_P} \sigma_{N_2}] \approx \delta_P$ hold. Therefore,

$$\delta > 0 \Leftrightarrow \exists \lambda \in \mathbb{N} \text{ s.t. } \mu_{N_1} + \alpha_{\delta_N} \sigma_{N_1} < \lambda < \mu_{N_2} - \alpha_{\delta_P} \sigma_{N_2}$$
$$\Leftrightarrow \mu_{N_1} + \alpha_{\delta_N} \sigma_{N_1} < \mu_{N_2} - \alpha_{\delta_P} \sigma_{N_2}$$

holds approximately. By solving $\mu_{N_1} + \alpha_{\delta_N} \sigma_{N_1} \approx \mu_{N_2} - \alpha_{\delta_P} \sigma_{N_2}$ we have the following approximate boundary $\delta \approx 0$ over N_1-N_2 plane:

$$(N_2 - N_1)(p_{\mathrm{TP}} + p_{\mathrm{TN}} - 1) \approx \alpha_{\delta_N} \sqrt{N_1 p_{\mathrm{TP}}(1 - p_{\mathrm{TP}}) + (K - N_1)p_{\mathrm{TN}}(1 - p_{\mathrm{TN}})}$$
$$+ \alpha_{\delta_P} \sqrt{N_2 p_{\mathrm{TP}}(1 - p_{\mathrm{TP}}) + (K - N_2)p_{\mathrm{TN}}(1 - p_{\mathrm{TN}})}. \tag{3}$$

In the case with $p_{\mathrm{TN}} = p_{\mathrm{TP}}$, the approximate boundary becomes a simple line:

$$N_2 \approx N_1 + \frac{(\alpha_{\delta_N} + \alpha_{\delta_P})\sqrt{K p_{\mathrm{TP}}(1 - p_{\mathrm{TP}})}}{2 p_{\mathrm{TP}} - 1} \tag{4}$$

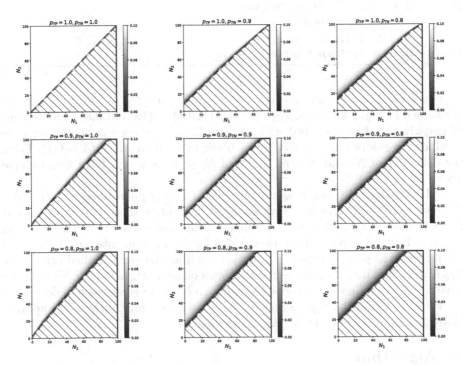

Fig. 1. δ-values at $(N_1, N_2) \in [0, K] \times [0, K]$ for $K = 100$ and $\delta_P = \delta_N = 0.1$ and $(p_{\mathrm{TP}}, p_{\mathrm{TN}}) \in \{1.0, 0.9, 0.8\} \times \{1.0, 0.9, 0.8\}$. $p_{\mathrm{TP}} = 1.0, 0.9, 0.8$ for top, middle and bottom graphs, respectively, and $p_{\mathrm{TN}} = 1.0, 0.9, 0.8$ for left, center and right graphs, respectively. Regions of $\delta < 0$ are filled with oblique lines. The regions of $\delta \geq 0$ are colored in grayscale from black ($\delta = 0$) to white ($\delta = 0.1$). The range of δ is $[-0.9, 0.1]$ because $\delta_P = \delta_N = 0.1$ in these experiments. The approximate boundary by the expression (3) is shown by a gray dashed line on each graph and it looks good approximations.

We give a graphical representation of δ-value over N_1-N_2 plane for fixed K, δ_P, δ_N and various p_{TP}, p_{TN} in Fig. 1.

---

noend 1. Algorithm for One-Threshold Classification Bandit Problem

---

Input: K: number of arms, θ: reward threshold, δ: confidence parameter,
      λ: threshold on the number of positive arms
Output: "positive" or "negative"
 1: initialization: $t \leftarrow 0$, $A \leftarrow \{1, 2, \dots, K\}$, $D \leftarrow [\,]$, $n_P, n_N \leftarrow 0$, $n_1, \dots, n_k \leftarrow 0$
 2: **loop**
 3:   $t \leftarrow t + 1$
 4:   $i_t \leftarrow \text{ASP}(D, K, \delta, \lambda, A)$ {ASP: Arm Selection Policy. See Sec. 4.2}
 5:   Pull arm i_t and update n_{i_t} as $n_{i_t} \leftarrow n_{i_t} + 1$
 6:   Get reward $X_{i_t}(n_{i_t})$ and append $(i_t, X_{i_t}(n_{i_t}))$ to D
 7:   Update $\overline{\mu}_{i_t}(t)$ and $\underline{\mu}_{i_t}(t)$ using Eqs. (5) and (6)
 8:   **if** $\overline{\mu}_{i_t}(t) < \theta$ **then**
 9:      $A \leftarrow A \setminus \{i_t\}$, $n_N \leftarrow n_N + 1$
10:      **if** $K - n_N < \lambda$ **return** "negative"
11:   **else if** $\underline{\mu}_{i_t}(t) \geq \theta$ **then**
12:      $A \leftarrow A \setminus \{i_t\}$, $n_P \leftarrow n_P + 1$
13:      **if** $n_P \geq \lambda$ **return** "positive"

For fixed $K = 100$ and $\delta_P = \delta_N = 0.1$, δ-values at $(N_1, N_2) \in [0, K] \times [0, K]$ are shown in Fig. 1 for $(p_{\text{TP}}, p_{\text{TN}}) \in \{1.0, 0.9, 0.8\} \times \{1.0, 0.9, 0.8\}$. The regions of negative δ-values are filled with oblique lines. The regions of non-negative δ-values are colored in grayscale from black ($\delta = 0$) to white ($\delta = 0.1$). From the definition of N_1 and N_2, the region of $N_2 \leq N_1$ is always filled with oblique lines.

When both of p_{TP} and p_{TN} are 1.0, δ is 0.1 at any point of region $N_1 < N_2$, because bad (good) arm is always assigned to positive (negative) arm. Even if there is no bad arm, $(1 - p_{\text{TN}})K$ arms are discriminated as positive arms on average. In fact, at $N_1 = 0$, we have $\delta \leq 0$ for N_2 in some interval $[0, N_2^0]$, and N_2^0 increases as p_{TN} decreases. Similarly, even if all the arms are bad arms, only $p_{\text{TP}}K$ arms are discriminated as positive arms on average. In fact, at $N_2 = K$, we have $\delta \leq 0$ for N_1 in some interval $[N_1^K, K]$, and N_1^K decreases as p_{TP} decreases.

We can see that the expression (3) gives good approximation from the approximate boundaries shown by a gray dashed lines.

4   Algorithm

The pseudocode of proposed algorithm for one-threshold classification bandits is shown in Algorithm 1. Note that the algorithm works for any arm selection policy.

4.1   Decision Condition

The upper and lower confidence bounds $\overline{\mu}_{i,n}$, $\underline{\mu}_{i,n}$ of μ_i after taking n samples from arm i are defined as follows:

$$\overline{\mu}_{i,n} = \hat{\mu}_{i,n} + \sqrt{\frac{1}{2n} \log \frac{2Kn^2}{\delta}}, \quad \underline{\mu}_{i,n} = \hat{\mu}_{i,n} - \sqrt{\frac{1}{2n} \log \frac{2Kn^2}{\delta}}, \qquad (5)$$

where $\hat{\mu}_{i,n}$ is the sample mean of rewards of arm i after n pulls. We use the following notations as well for the sake of simplicity.

$$\overline{\mu}_i(t) = \overline{\mu}_{i,n_i(t+1)}, \quad \underline{\mu}_i(t) = \underline{\mu}_{i,n_i(t+1)}. \tag{6}$$

Here, $n_i(t+1)$ is the number of pulls of arm i after t pulls in total.

The decision condition for positiveness of arm i is $\underline{\mu}_i(t) \geq \theta$ and that for negativeness of arm i is $\overline{\mu}_i(t) < \theta$.

For these decision conditions, the following lemma guarantees the probability of wrong decision for each arm is at most δ/K.

Lemma 2. *For a positive arm i (i.e., $\mu_i \geq \theta$), $\overline{\mu}_i(t) \geq \theta$ holds for any time step t with probability at least $1 - \frac{\delta}{K}$. For a negative arm i (i.e., $\mu_i < \theta$), $\underline{\mu}_i(t) < \theta$ holds for any time step t with probability at least $1 - \frac{\delta}{K}$.*

Proof. For a positive arm i, we have

$$\mathbb{P}[\exists t, \, \overline{\mu}_i(t) < \theta] = \mathbb{P}[\exists n, \, \overline{\mu}_{i,n} < \theta] \leq \sum_{n=1}^{\infty} \mathbb{P}[\overline{\mu}_{i,n} < \theta]$$

$$= \sum_{n=1}^{\infty} \mathbb{P}\left[\hat{\mu}_{i,n} + \sqrt{\frac{1}{2n} \log \frac{2Kn^2}{\delta}} < \theta\right]$$

$$\leq \sum_{n=1}^{\infty} \mathbb{P}\left[\hat{\mu}_{i,n} < \mu_i - \sqrt{\frac{1}{2n} \log \frac{2Kn^2}{\delta}}\right] \quad (\text{because } \mu_i \geq \theta)$$

$$\leq \sum_{n=1}^{\infty} \exp\left(-2n\left(\sqrt{\frac{1}{2n} \log \frac{2Kn^2}{\delta}}\right)^2\right) \begin{pmatrix} \text{by Hoeffding's} \\ \text{inequality} \end{pmatrix}$$

$$= \sum_{n=1}^{\infty} \frac{\delta}{2Kn^2} < \frac{\delta}{K} \left(\text{because } \sum_{n=1}^{\infty} \frac{1}{n^2} = \frac{\pi^2}{6} < 2\right).$$

Therefore, for a positive arm i, the probability that $\overline{\mu}_i(t) > \theta$ always holds for any time step t is larger than $1 - \frac{\delta}{K}$.

Similarly, we can show the inequality for a negative arm as well. $\square$

As $n \to \infty$, $\hat{\mu}_{i,n}$ goes to μ_i from law of large numbers and $\sqrt{\frac{1}{2n} \log \frac{2Kn^2}{\delta}}$ goes to 0. Therefore, a positive arm i satisfies $\underline{\mu}_{i,n} > \theta$ (i.e. it is diagnosed as a positive arm) for some finite n if $\mu_i \neq \theta$. Here $\mu_i = \theta$ generally corresponds to the situation that infinite number of draw of the arm i is required for diagnosis. From Lemma 2, the probability that a positive arm i satisfies $\overline{\mu}_i(t_0) < \theta$ (i.e. it is diagnosed as a negative arm) is at most $\frac{\delta}{K}$. Therefore a positive arm is diagnosed as a positive arm correctly with failure probability at most $\frac{\delta}{K}$. Similarly, a negative arm is diagnosed correctly with failure probability at most $\frac{\delta}{K}$. Therefore the failure probability that the agent diagnoses any arm wrongly is at most δ as long as $\mu_i \neq \theta$ for any arm i. The agent counts the number of positive arm n_P and negative arm n_N by using these conditions at each time step t, and stops and outputs "positive" when $n_P \geq \lambda$ or outputs "negative" when $K - n_N < \lambda$ since the number of positive arms cannot exceed $K - n_N$. Since the counts n_P and n_N are correct with probability at least $1 - \delta$ from the above discussion, Algorithm 1 solves one-threshold classification bandits with a specified confidence. We have proved the following theorem.

Theorem 2. *For one-threshold classification bandits with parameters $K, \theta, \lambda, \delta$, the outputs of Algorithm 1 using any arm selection policy satisfy the requirements of Problem 2.*

4.2 Arm Selection Policies

Let A_t be a set of arms that have not been diagnosed as positive or negative before time step t by the conditions explained in the previous section. It is enough to choose arm only from A_t at each time step t.

We developed the arm selection policy based on the Thompson sampling. Let θ_i be the parameter of the reward distribution ν_i of arm i. Assume a prior distribution π_i^0 of θ_i. The original Thompson sampling estimates a posterior distribution π_i^{t-1} of θ_i for each arm i at each time step t, and chooses an arm. The proposed algorithm is described as follows:

ThompsonSampling-CB:
1. For each arm $i \in A_t$,
(1) Calculate the posterior distribution π_i^{t-1} of θ_i using all the rewards obtained by time step t.
(2) Sample $\hat{\theta}_i \sim \pi_i^{t-1}$.
(3) Calculate the expected mean for given $\hat{\theta}_i$: $\tilde{\mu}_i^t = \mathbb{E}_{P[X|i,\hat{\theta}_i]}[X]$.
2. Count the number of arms i with $\tilde{\mu}_i^t$ at least θ, $B_t = |\{i \in [K] | \tilde{\mu}_i^t \geq \theta\}|$.
3. Select arm $i_t = \begin{cases} \underset{i \in A_t}{\arg\max}\, \tilde{\mu}_i^t & \text{(when } B_t \geq \lambda) \\ \underset{i \in A_t}{\arg\min}\, \tilde{\mu}_i^t & \text{(when } B_t < \lambda) \end{cases}$

For comparison, we examine the following arm selection policies as well.

UCB-CB:
Select $i_t = \underset{i \in A_t}{\arg\max}\, \hat{\mu}_i(t) + \sqrt{\frac{1}{2n_i(t)} \log t}$

Successive Elimination-CB:
Select $i_t = \underset{i \in A_t}{\arg\min}\, n_i(t)$

In both arm selection policies, if more than one argument satisfy the condition, one of them is chosen arbitrarily. As one of comparison methods, we select UCB algorithm because it is the best performer for good arm identification problem [4], whose problem setting is most similar to the setting of our one-threshold classification problem.

5 Experiments

In this section, we show the result of comparison experiments for the three algorithms that we proposed in Sect. 4.2.

The stopping time of Algorithm 1 using ThompsonSampling-CB is compared with those using UCB-CB and Successive Elimination-CB for the one-threshold classification bandits with positive δ that is reduced from classification bandits instances. In this experiment, we fixed parameters $K, \theta, p_{\mathrm{TN}}, p_{\mathrm{TP}}, \delta_P$ and δ_N of the original classification bandits instances as $K = 20$, $\theta = 0.5$, $p_{\mathrm{TN}} = p_{\mathrm{TP}} = 0.95$ and $\delta_P = \delta_N = 0.1$. Expected reward μ_i of arm i is taken from a uniform distribution over $[0, \theta]$ for negative arms and $[\theta, 1]$ for positive arms. The distributions of reward are Bernoulli distribution. For $N_1 = 0, 1, \ldots, 14$, $N_2 = N_1 + 5$ and the cases with just $5, 10, 15$ positive arms, we reduced each problem instance to the corresponding one-threshold problem with parameters λ and δ, and measured the stopping time (the number of samples) of Algorithm 1.

The results are shown in the upper left, the upper right and the lower left graphs for the cases with just $5, 10, 15$ positive arms, respectively, of Fig. 2. For

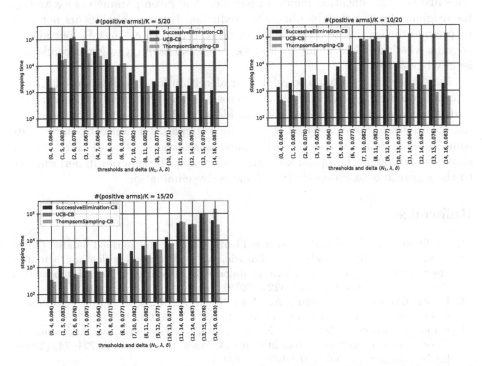

Fig. 2. Average stopping times (the number of samples) over 100 runs of Algorithm 1 using three arm selection policies for one-threshold classification bandits instances with parameters λ and δ reduced from classification bandits instances with parameters $K = 20$, $\theta = 0.5$, $p_{\mathrm{TN}} = p_{\mathrm{TP}} = 0.95$, $\delta_P = \delta_N = 0.1$, $N_1 = 0, 1, \ldots, 14$ and $N_2 = N_1 + 5$. y-axis is 'log' scale. The upper left, the upper right and the lower left graphs are results for the case with just $5, 10$, and 15 positive arms, respectively.

these three graphs, we can see ThompsonSampling-CB always stops earlier than SuccessiveElimination-CB and always stops earlier or in time comparable to UCB-CB. The performance of UCB-CB is poor when the output should be "negative" because UCB-CB tries to select positive arms with higher priority.

6   Conclusion and Future Works

In this paper, we presented an algorithm to reduce classification bandits problem based on an imperfect classifier with nonnegligible false-positive and false-negative rates into a one-threshold classification bandits problem under the allowed error rate δ. The parameters of true negative and positive probabilities p_{TN}, p_{TP}, and the number of arms K are supposed to be given in actual applications. Then the question here was whether we can still discriminate the number of bad arms is at most N_1 with probability at least $1 - \delta_N$ or at least $N_2 (> N_1)$ with probability at least $1 - \delta_P$. Usually confidence parameter δ required for bandits algorithms is the same as a given parameter itself, but in classification bandits, the confidence parameter δ required for the transformed one-threshold classification bandits is smaller than given parameters δ_N and δ_P for original classification bandits. Our reduction theorem enables us not only to provide the error rate δ smaller than originally given δ_P and δ_N but also to suggest whether it is difficult to find algorithm satisfying the given confidence level when $\delta < 0$.

For future work, we plan to apply our algorithm for classification bandits to interactive measurement by Raman microscope for differentiating cancer cells and non-cancer cells, where no one can identify whether each cell is cancer or not with 100% accuracy (at best 80–95% for example). Theoretically there exists some room to improve the algorithm such as selection policy although in our simulation through Thompson sampling was found to be superior in performance to the algorithms based on UCB and successive elimination.

References

1. Audibert, J.Y., Bubeck, S.: Best arm identification in multi-armed bandits (2010)
2. Gabillon, V., Ghavamzadeh, M., Lazaric, A.: Best arm identification: a unified approach to fixed budget and fixed confidence. In: Advances in Neural Information Processing Systems, pp. 3212–3220 (2012)
3. Kalyanakrishnan, S., Tewari, A., Auer, P., Stone, P.: Pac subset selection in stochastic multi-armed bandits. In: ICML, vol. 12, pp. 655–662 (2012)
4. Kano, H., Honda, J., Sakamaki, K., Matsuura, K., Nakamura, A., Sugiyama, M.: Good arm identification via bandit feedback. Mach. Learn. **108**(5), 721–745 (2019). https://doi.org/10.1007/s10994-019-05784-4
5. Kaufmann, E., Koolen, W.M., Garivier, A.: Sequential test for the lowest mean: from Thompson to murphy sampling. In: Advances in Neural Information Processing Systems, pp. 6332–6342 (2018)
6. Locatelli, A., Gutzeit, M., Carpentier, A.: An optimal algorithm for the thresholding bandit problem. In: Proceedings of the 33rd International Conference on International Conference on Machine Learning, vol. 48, pp. 1690–1698 (2016)

7. Nachlas, J.A., Loney, S.R., Binney, B.A.: Diagnostic-strategy selection for series systems. IEEE Trans. Reliab. **39**(3), 273–280 (1990)
8. Pelissier, A., et al.: Intelligent measurement analysis on single cell Raman images for the diagnosis of follicular thyroid carcinoma. arXiv preprint (2019). arxiv.org/abs/1904.05675
9. Raghavan, V., Shakeri, M., Pattipati, K.: Test sequencing algorithms with unreliable tests. IEEE Trans. Syst. Man Cybern.-Part A: Syst. Humans **29**(4), 347–357 (1999)
10. Tabata, K., Nakamura, A., Honda, J., Komatsuzaki, T.: A bad arm existence checking problem: how to utilize asymmetric problem structure? Mach. Learn. **109**, 1–46 (2019). https://doi.org/10.1007/s10994-019-05854-7

The First Workshop and Shared Task on Scope Detection of the Peer Review Articles (SDPRA 2021)

Overview and Insights from Scope Detection of the Peer Review Articles Shared Tasks 2021

Saichethan Miriyala Reddy[1] and Naveen Saini[2(✉)]

[1] Indian Institute of Information Technology Bhagalpur, Bhagalpur, Bihar, India
miriyala.cse.1725@iiitbh.ac.in
[2] Université Toulouse III - Paul Sabatier, IRIT, UMR 5505 CNRS, Toulouse, France
Naveen.Saini@irit.fr

Abstract. In the current paper, we will present the results of our shared task at *The First Workshop & Shared Task on Scope Detection of the Peer Review Articles* (SDPRA) collocated with PAKDD 2021. It aims to develop system(s) which can help in the peer-review process in the initial screening usually performed by the editor(s). We received four submissions in total: three from academic institutions and one from the industry. The quality of submission shows a greater interest in the task by the research community.

1 Introduction

In recent years, scientific articles are published at a rapid rate and can be accessed using search capability of scholarly search engines like Google Scholar[1], dblp[2], among others. This continuous growth help the researchers in getting the recent developments in the respective domain [3] as well as discovering new research dire. But before publishing, peer review is the widely accepted method for the validation of the submitted papers to the conferences/journals. In the academic peer review process, the first stage begins at the editor's desk where task is to identify the in-scope and out-of-scope articles and then reject out-of-scope submissions or in other words, it is the job of the editor, who is also an expert in the particular field to take decisions whether an article should be rejected without further review or forwarded to expert reviewers for meticulous evaluation. Some of the common reasons for rejection are due to paper's language and writing/formatting style, results are not better than state-of-the-art, method is too simple (seriously? Isn't that a good thing?), too narrow or outdated or out of scope, among others.

Without any automatic system, the Editor may spends a substantial amount of time in finding the appropriateness of the submitted article or before passing it to the reviewers for review purpose. Inspite of having good quality of the submitted articles, many articles got rejection because of their out-of-score [18].

---

[1] https://scholar.google.com/.
[2] https://dblp.uni-trier.de/.

© Springer Nature Switzerland AG 2021
M. Gupta and G. Ramakrishnan (Eds.): PAKDD 2021 Workshops, LNAI 12705, pp. 73–78, 2021.
https://doi.org/10.1007/978-3-030-75015-2_7

Machine learning and artificial intelligence methods make it possible to identify the in-scope and out-of-scope of the scientific publications. However, in order to improve the performance of these methods and to carry out the experiments, researchers needs to access and use of large database of scientific publications. This shared task aims to bring undergraduate and master students, NLP/ML researchers and from other backgrounds who: (a) have a deep interest in mining scientific articles; (b) develop novel methods that able to improve the performance. The purpose of the shared task is to provide a platform for developing and evaluating such models. To this end, the shared task provides a dataset covering a broad range of topics within the computer science. The importance of the shared task can be easily understood from the recent papers [6,11,17].

In this paper, we present the task description, dataset, discussion about the participating systems followed by their results and insight from shared task.

2   Problem Definition and Datasets

This shared task focuses on identifying topics or category of scientific articles, which helps in pre-screening of the scientific articles at the Editor's desk to decide whether the article is out-of-score or not. For this task we collected a total of 35000 abstracts of the scientific article in computer science domain from ARXIV. Given an abstract of a paper, the objective of the shared task is to classify it into one of the 7 predefined categories listed in Table 1. The dataset is divided into training, validation and testing set. The statistics about the datasets is shown in Fig. 1. We have openly released our datasets at http://dx.doi.org/10.17632/njb74czv49.1 as a part of Mendeley data [16].

Table 1. Pre-defined categories for scientific article classification and their description. Here, Abb. means abbreviation of the categories used in this paper.

Category Name	Abb.	Description of the category
Computation & Language(CL)	CL	It can be defined as an application of computer science to the analysis, synthesis and comprehension of written and spoken language.
Cryptography & Security(CR)	CR	In this category, secure communication related articles are included.
Distributed & Cluster Computing	DC	It refers to the application of cluster and distributed computing systems in order to leverage processing power, memory, etc., of any computing system for an increase in efficiency.
Data Structures & Algorithms	DS	It refers to the study of using an entity that stores and organizes data in order to develop algorithms that help to decode and optimize computer programs.
Logic in Computer Science	LO	It refers to the overlap between the field of logic and that of computer science.
Networking & Internet Architecture	NI	It refers to the specification of a network's physical components and their functional organization and configuration.
Software Engineering	SE	It is the systematic application of engineering approaches to the development of software.

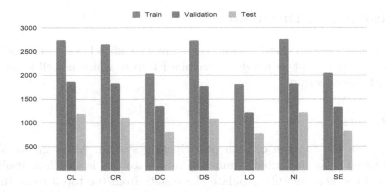

Fig. 1. Distribution of data

3   Participating Teams

For this shared task, 18 teams have registered, but only 4 were able to submit their systems and were able complete the final evaluation correctly. Description about the participated team and their methods are brief described below. The systems of all the teams are based on supervised techniques.

- **UTMN:** This team was from University of Tyumen, Russia. The method [8] developed by this team uses the SciBERT language model [1] along with topic modeling tool (Latent Dirichlet Allocation [2]) to obtain the good quality of predictions. Several other language models like RoBERTa [13], PubMedBERT [9], Topic-informed BERT-based models (tBERT) [14], among others, are also investigated to analyze the performance.
- **IIITT:** This system was submitted by the team members from Indian Institute of Information Technology Tiruchirapalli, India. The authors [10] utilize the transfer learning-based approach for the scope identification of the scientific articles. They fine-tuned the existing model namely, BERT, RoBERTa, and SCIBERT using the provided datasets and then, tested on the test dataset. Their results shows that ensemble approach over the three models perform the best.
- **parklize:** The author of this method [15] was from Maynooth University, Ireland and uses sentence embedding obtained using pre-trained SciBERT model along with entity embeddings mentioned in the text. The entities in the abstract are tagged using the *TagMe* model [5] and then, wikipedia2vec tool [19] is used to get the embeddings of the entities. For classification, seven classifiers are used followed by majority-voting ensemble approach.
- **FideLIPI:** The system by Ghosh et al. [7] developed four different sub-systems for the shared task and ensemble the predictions of these four sub-systems. The four sub-systems includes: (a) Abstract (of the scientific articles) level RoBERTa; (b) RoBERTa based model with additional features extracted using Latent Dirichlet Allocation [2]; (c) Sentence level RoBERTa model, and (d) Logistic regression [12] model utilizing features provided by tf-idf model [4].

4   Results and Discussion

As our task deals with classification; therefore, the official evaluation metric for the shared task is kept as weighted-average F1 score which is well-know in the field of Machine learning.

4.1   Results

Table 2 show the top run of each participating system. Although we allowed teams to submit an unlimited number of runs since this is an offline evaluation on a blind test set, we only tabulate the results from the top 3 runs. In order to compare between the systems, we considered the $F1$ scores. The ranking of systems based on their performance is also shown in the first column of Table 2. It has been observed that the system *UTMN* obtained the first rank and *FideLIPI* obtained the second rank. All teams have utilized transformer-based language models to develop their systems.

All systems shown to have a good performance. We have noticed that high classification performance of the abstracts are due to the use of single domain like computer science and also due to creating the test set from same sources as development and training. From a language perspective, most of the systems utilized transformer based language models, some even used pre-trained models on scientific corpora. However, we believe that more efforts should be focused developing computationally simpler models. In the future we plan to extend this corpus to different discipline & domains like history, astro-physics, etc., with the hope that SDPRA will help foster further research in this domain.

Table 2. The results reported by different submitted systems and their ranking

Rank	Team name	F1 Score		
		Run 1	Run 2	Run 3
1	UTMN	0.9370	0.9354	**0.9382**
2	FideLIPI	**0.9293**	0.9122	0.9163
3	IIITT	0.9227	0.9156	**0.9246**
4	Parklize	0.9173	**0.9206**	0.9151

4.2   Code Reproducibility

To improve code reproducibility and transparency in scientific community, all shared task participants are asked to submit their systems to our Github Repository[3]. The codes are open to use for the scientific community.

---

[3] https://github.com/SDPRA-2021/shared-task.

5   Conclusion and Future Scope

The first shared task on 'Scope Detection of the Peer Review Articles' comprises identification of *in-scope* and *out-of-scope* categories using only the abstract of the *arxiv* articles. We received four submissions. The reported high performance by different systems shows the high coherence among the abstracts, which needs expansion by including the abstracts of the other domains. Moreover, most of the submitted systems were based on utilizing a deep learning-based model, but we are interested in an elementary and unsupervised model. We are also interested in collaborating with NLP and AI researchers to interact and discuss the new tools to be developed for the scope identification.

Acknowledgment. We would like to thanks all steering committee members, Pushpak Bhattacharyya (IIT Bombay), Antoine Doucet (Universite de La Rochelle), Sriparna Saha (IIT Patna), Jose G Moreno (Universite de Toulouse, IRIT), Adam Jatowt (Kyoto University), Anita de Waard (Elsevier), Gael Dias (University of Caen Normandie) for their support. We are also thankful to all the members of our program committee for their time and valuable inputs.

References

1. Beltagy, I., Lo, K., Cohan, A.: SciBERT: a pretrained language model for scientific text. arXiv preprint arXiv:1903.10676 (2019)
2. Blei, D.M., Ng, A.Y., Jordan, M.I.: Latent Dirichlet allocation. J. Mach. Learn. Res. **3**, 993–1022 (2003)
3. Cohan, A., Goharian, N.: Scientific document summarization via citation contextualization and scientific discourse. Int. J. Digit. Libr. **19**(2–3), 287–303 (2018). https://doi.org/10.1007/s00799-017-0216-8
4. Fautsch, C., Savoy, J.: Adapting the TF IDF vector-space model to domain specific information retrieval. In: Proceedings of the 2010 ACM Symposium on Applied Computing, pp. 1708–1712 (2010)
5. Ferragina, P., Scaiella, U.: TAGME: on-the-fly annotation of short text fragments (by wikipedia entities). In: Proceedings of the 19th ACM International Conference on Information and Knowledge Management, pp. 1625–1628 (2010)
6. Ghosal, T., Verma, R., Ekbal, A., Saha, S., Bhattacharyya, P.: Investigating impact features in editorial pre-screening of research papers. In: Proceedings of the 18th ACM/IEEE on Joint Conference on Digital Libraries, pp. 333–334 (2018)
7. Ghosh, S., Chopra, A.: Using transformer based ensemble learning to classify scientific articles. In: Gupta, M., Ramakrishnan, G. (eds.) PAKDD 2021. LNAI, vol. 12705, pp. 106–113. Springer, Heidelberg (2021)
8. Glazkova, A.: Identifying topics of scientific articles with BERT-based approaches and topic modeling. In: Gupta, M., Ramakrishnan, G. (eds.) PAKDD 2021. LNAI, vol. 12705, pp. 98–105. Springer, Heidelberg (2021)
9. Gu, Y., et al.: Domain-specific language model pretraining for biomedical natural language processing. arXiv preprint arXiv:2007.15779 (2020)
10. Hande, A., Puranik, K., Priyadharshini, R., Chakravarthi, B.R.: Domain identification of scientific articles using transfer learning and ensembles. In: Gupta, M., Ramakrishnan, G. (eds.) PAKDD 2021. LNAI, vol. 12705, pp. 88–97. Springer, Heidelberg (2021)

11. Kelly, J., Sadeghieh, T., Adeli, K.: Peer review in scientific publications: benefits, critiques, & a survival guide. Ejifcc **25**(3), 227 (2014)
12. Kleinbaum, D.G., Dietz, K., Gail, M., Klein, M., Klein, M.: Logistic Regression. Springer, Heidelberg (2002). https://doi.org/10.1007/b97379
13. Liu, Y., et al.: Roberta: a robustly optimized BERT pretraining approach. arXiv preprint arXiv:1907.11692 (2019)
14. Peinelt, N., Nguyen, D., Liakata, M.: tBERT: topic models and BERT joining forces for semantic similarity detection. In: Proceedings of the 58th Annual Meeting of the Association for Computational Linguistics, pp. 7047–7055 (2020)
15. Piao, G.: Scholarly text classification with sentence BERT and entity embeddings. In: Gupta, M., Ramakrishnan, G. (eds.) PAKDD 2021. LNAI, vol. 12705, pp. 79–87. Springer, Heidelberg (2021)
16. Reddy, S., Saini, N.: SDPRA 2021 shared task data V1 (2021). https://doi.org/10.17632/njb74czv49.1
17. Sergeeva, E., Zhu, H., Prinsen, P., Tahmasebi, A.: Negation scope detection in clinical notes and scientific abstracts: a feature-enriched LSTM-based approach. In: AMIA Summits on Translational Science Proceedings 2019, p. 212 (2019)
18. Trumbore, S., Carr, M.E., Mikaloff-Fletcher, S.: Criteria for rejection of papers without review (2015)
19. Yamada, I., et al.: Wikipedia2vec: an efficient toolkit for learning and visualizing the embeddings of words and entities from wikipedia. arXiv preprint arXiv:1812.06280 (2018)

Scholarly Text Classification with Sentence BERT and Entity Embeddings

Guangyuan Piao[✉] [iD]

Department of Computer Science, Hamilton Institute, Maynooth University,
Maynooth, Co Kildare, Ireland
guangyuan.piao@mu.ie

Abstract. This paper summarizes our participated solution for the shared task of the text classification (scope detection) of peer review articles at the SDPRA (Scope Detection of the Peer Review Articles) workshop at PAKDD 2021. By participating this challenge, we are particularly interested in how well those pre-trained word embeddings from different neural models, specifically transformer models, such as BERT, perform on this text classification task. Additionally, we are also interested in whether utilizing entity embeddings can further improve the classification performance. Our main finding is that using SciBERT for obtaining sentence embeddings for this task provides the best performance as an individual model compared to other approaches. In addition, using sentence embeddings with entity embeddings for those entities mentioned in each text can further improve a classifier's performance. Finally, a hard-voting ensemble approach with seven classifiers achieves over 92% accuracy on our local test set as well as the final one released by the organizers of the task. The source code is publicly available at https://github.com/parklize/pakdd2021-SDPRA-sharedtask.

Keywords: Text classification · Word embeddings · Sentence embeddings · Entity embeddings

1 Introduction

Text classification is a crucial task in Natural Language Processing (NLP) with a wide range of applications such as scope detection of scholarly articles [5], sentiment classification [7], and news topic classification [11]. For example, identifying the topics or category of scientific articles, can help us efficiently manage a large number of articles, retrieve related papers, and build personal recommendation systems. In this report, we present the details of our solution for scholarly text (abstract) classification in the context of a shared task at the SDPRA (Scope Detection of the Peer Review Articles) workshop [8][1] at PAKDD 2021. By participating this task, we are particularly interested in understanding (1) how good

---

[1] https://sdpra-2021.github.io/website/.

© Springer Nature Switzerland AG 2021
M. Gupta and G. Ramakrishnan (Eds.): PAKDD 2021 Workshops, LNAI 12705, pp. 79–87, 2021.
https://doi.org/10.1007/978-3-030-75015-2_8

the transfer learning performance is based on pre-trained transformer networks designed for sentence/text embeddings, and (2) whether entity embeddings for those mentioned entities in a text help to improve the classification performance.

1.1 Task Description

The shared task is a multi-class classification problem, which aims to classify each given abstract of a scholarly article into one of seven classes. Overall, there are 35,000 abstracts in total where 16,800 for training, 11,200 for validation, and 7,000 for final testing [9]. Table 1 shows the distribution of each class of the dataset.

For final submission, each participated team allows to submit results with three different runs in which the best-performing result will be chosen for the final ranking. The official evaluation metric for the shared task is *weighted-average F1 score*[2].

Table 1. Dataset statistics.

Class	Train	Validation	Test
Computation and Language (CL)	2,740	1,866	1,194
Cryptography and Security (CR)	2,660	1,835	1,105
Distributed and Cluster Computing (DC)	2,042	1,355	803
Data Structures and Algorithms (DS)	2,737	1,774	1,089
Logic in Computer Science (LO)	1,811	1,217	772
Networking and Internet Architecture (NI)	2,764	1,826	1,210
Software Engineering (SE)	2,046	1,327	827
Total	**16,800**	**11,200**	**7,000**

2   Proposed Approach

Figure 1 illustrates a simplified architecture of a base model where a set of base models can be used for an ensemble approach. The key intuition is that maximizing transfer learning via pre-trained sentence encoders available on the Web so that we can obtain text (abstract) embeddings via those models in a straightforward manner. Afterwards, we can train our separate FNNs (Feed-forward Neural Networks) on top of those text embeddings for our classification task.

2.1   Pre-trained Sentence Encoders

In the following, we briefly introduce different pre-trained sentence encoders used in our approach.

---

[2] https://scikit-learn.org/stable/modules/generated/sklearn.metrics.f1_score.html.

Sentence Transformers. Sentence-BERT [10] is a modification of the BERT [3] network using siamese and triplet networks that are able to derive semantically meaningful sentence embeddings. SentenceTransformers[3] is a Python framework for state-of-the-art sentence and text embeddings. The models in SentenceTransformers are based on transformer networks like BERT and RoBERTa, and to facilitate transfer learning, the framework also provides a wide range of pre-trained sentence transformers which can be used for encoding a sentence/text. Therefore, we tested and used some of the pre-trained sentence transformers to encode each abstract, and used those encoded sentences (embeddings) as an input to additional FNNs for our task. The list of used model names can be found in Table 3.

SciBERT [1]. Although the pre-trained models in SentenceTransformers can be directly used for obtaining abstract embeddings for our task, they are trained on general domain corpora such as news articles and Wikipedia, which might have some limitations for the obtained embeddings as the domain of our task is the Computer Science domain. In contrast to those aforementioned sentence transformers, SciBERT is trained on papers from the corpus of the Semantic Scholar[4] which contains 1.14 million papers with 3.1 billion tokens using the full text of the papers for training.

Universal Sentence Encoder [2]. Similar to sentence transformers, this approach also provides the functionality for encoding sentences into corresponding embeddings that specifically target transfer learning to other NLP tasks. The model (`universal-sentence-encoder/4` in Table 3) is available in Tensorflow Hub[5], which is a repository of trained machine learning models ready for fine-tuning and deployable anywhere.

Encoding Text with Entity Embeddings. Motivated by the recent studies, which have shown that leveraging entities mentioned in a short text improves text classification performance [11,12], we obtain text embeddings via the set of entities mentioned in them. There are four main steps in this process.

1. Wikipedia entities mentioned in a text are extracted using TagMe [4]. Figure 2 shows the set of extracted Wikipedia entities using TagMe from a given abstract. In addition to an entity, TagMe also provides its *confidence score*, e.g., `Graph (discrete mathematics)`:0.43.
2. For each entity, we obtain pre-trained corresponding embeddings provided from wikipedia2vec [14]. wikipedia2vec is a tool used for obtaining embeddings of words and entities from Wikipedia, and also provides pre-trained word and entity embeddings[6].

---

[3] https://www.sbert.net/.
[4] `semanticscholar.org`.
[5] https://www.tensorflow.org/hub.
[6] https://wikipedia2vec.github.io/wikipedia2vec/pretrained/.

3. Despite of the efficiency of TagMe for extracting mentioned entities from a text, some of those extracted entities can be noisy. To cope with this problem, we further apply k-means clustering approach for those extracted entity embeddings with two clusters - one relevant cluster with the majority of related entities for a given text and the other cluster which is noisy - with the assumption that the larger one should contain the most of high-quality entities. Therefore, we only consider the entities belong to the larger cluster and do not consider those in the other cluster.

4. Finally, a text embedding can be obtained with the set of entity embeddings for those entities mentioned in the text (abstract) via a weighted mean pooling. That is, the entity embeddings are weight averaged based on their confidence scores where those scores can be treated as attention weights [13].

We use `entity-emb` to indicate this encoding approach for deriving a text embedding from a raw text. As we can see from Fig. 1, the text embedding obtained with `entity-emb` can be optionally concatenated together with another text embedding obtained via a pre-trained sentence encoder such as SciBERT as an input to FNNs for classification.

fastText. The base models with pre-trained sentence encoders do not allow fine tuning of word embeddings in the context of our task. To better understand whether using those pre-trained sentence encoders provide good transfer learning quality, we also train a fastText [6][7] text classifier, which is a fast and efficient text classifier from Facebook using "bag of" tricks. It is often on par with deep learning classifiers in terms of accuracy, but with many orders of magnitude faster for training and evaluation.

Hard-voting ensemble. For the final classification with the predictions of multiple base models, we adopt the hard-voting ensemble. It sums the votes from those models, and then, the class with the most votes is used as the predicted class.

2.2   Training

We train those base models introduced in Sect. 2.1 using the training set provided by the challenge organizers with the objective of maximizing the evaluation

Table 2. Dataset statistics where the validation set (11,200) is further split into internal validation and test sets evenly.

Train	Internal validation	Internal test	Final test
16,800	5,600	5,600	7,000

---

[7] https://fasttext.cc/.

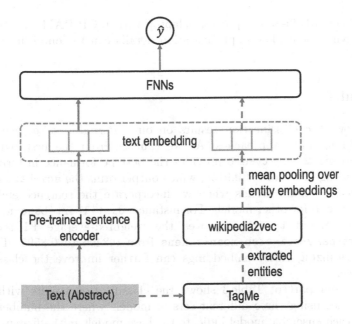

Fig. 1. Simplified illustration of a base model for encoding text into a text embedding for classification. The components connected via dashed lines are optional.

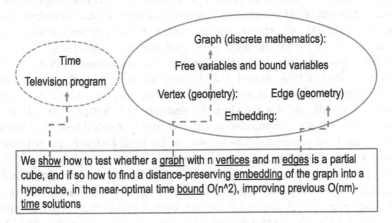

Fig. 2. Example of applying k-means clustering for the extracted entities of a text with $k = 2$. Those entities in the larger cluster–the green one–are chosen while those in the other cluster are not used. (Color figure online)

metric: weighted-average F1 score. We further divide the validation set into *internal* validation and test sets where each set contains 5,600 examples as shown in Table 2. The internal validation set is used for hyper-parameter tuning if needed for some base models, and the internal test set is used for testing only for different base models and ensemble approaches. All experiments are run on an

Intel(R) Core(TM) i5-8365U processor laptop with 16GB RAM and the Google Colab [8] environment. The implementation details can be found in our github repository[9].

3   Results

Table 3 shows the classification results on our *internal* test set using different transfer learning approaches. As we can see from the first part of the table, `sci-bert & entity-emb` provides the best performance in terms of the weighted-average F1 score of 0.9158, which outperforms the `sci-bert` (0.9140). We also notice similar trends when we incorporate the sentence embeddings with `entity-emb` for other models. For instance, `distilbert-base-nli-stsb-mean-tokens & entity-emb` improves the weighted-average F1 score over `distilbert-base-nli-stsb-mean-tokens` from 0.8496 to 0.8595. This indicates that utilizing entity embeddings can further improve the classification performance.

The bottom part of Table 3 shows the classification results with ensemble approaches using three different sets of models where the numbers in the bracket of each ensemble model indicate the base models in the first part of the table. For instance, `ensemble[4,5,7,8,9]` indicates the (hard-voting) ensemble of 4th, 5th, 7th, 8th, and 9th models in Table 3, i.e., `distilbert-base-nli-stsb-mean-tokens`, `distilroberta-base-msmarco-v2`, `universal-sentence-encoder/4`, `fasttext`, `sci-bert`, and `sci-bert & entity-emb`, respectively. As we can see from the table, the hard voting scheme with the classification results from different base models can further improve the performance. Note that incorporating the weakest base model `distilbert-base-nli-stsb-quora-ranking` also helped improve the performance as we can see from the three ensemble approaches. Overall, we observe that the `ensemble[4,5,6,7,8,9,10]` provides the best classification performance on our *internal* test set. These three ensemble models are used for predicting the labels on the *final* test set and submission.

Figure 3 further illustrates the confusion matrix using `ensemble[4,5,6,7,8,9,10]` on our *internal* test set. Each cell in the confusion matrix indicates the number of samples predicted as the corresponding column label which have the true row label. For example, 570 in the top-left cell indicates 570 samples out of 5,600 samples (10.8%) in the *internal* test set have been classified as LO where LO is the true label. The first cell in the second row shows that 4 (0.07% of the 5,600) samples have been classified as LO using `ensemble[4,5,6,7,8,9,10]` where the true label is DC. As we can see from the figure, DS (Data Structures and Algorithms) and DC (Distributed and Cluster Computing) are the most confusing labels followed by NI (Networking and Internet Architecture) and DC. The following abstract shows an example with the ground truth label as DC and with the predicted label as DS by our approach.

---

[8] https://colab.research.google.com/.
[9] https://github.com/parklize/pakdd2021-SDPRA-sharedtask.

Table 3. Classification results on the *internal* test set.

No.	Model name	Weighted-average F1 score
0	distilbert-base-nli-mean-tokens	0.8389
1	distilroberta-base-paraphrase-v1	0.8641
2	xlm-r-distilroberta-base-paraphrase-v1	0.8625
3	roberta-base-nli-stsb-mean-tokens	0.8413
4	distilbert-base-nli-stsb-mean-tokens	0.8496
5	distilroberta-base-msmarco-v2	0.8591
6	distilbert-base-nli-stsb-quora-ranking	0.8228
7	universal-sentence-encoder/4	0.8931
8	fasttext	0.8816
9	sci-bert	0.9140
10	sci-bert & entity-emb	**0.9158**
11	ensemble[4,5,7,8,9]	0.9201
12	ensemble[4,5,7,8,9,10]	0.9252
13	ensemble[4,5,6,7,8,9,10]	**0.9258**

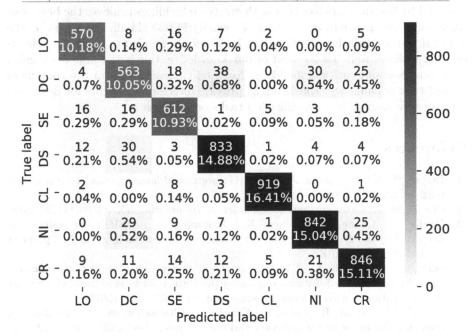

Fig. 3. The confusion matrix of classification using ensemble[4,5,6,7,8,9,10] on the *internal* test set (with 5,600 samples). Each cell contains the number of samples and its corresponding percentage of the 5,600 samples for the corresponding predicted and true labels.

"We present DegreeSketch, a semi-streaming distributed sketch data structure and demonstrate its utility for estimating local neighborhood sizes and local triangle count heavy hitters on massive graphs. DegreeSketch consists of ..."

As we can see from the example, this abstract is about an article proposing a data structure in the context of distributed computing, which is difficult to classify since the label DS also makes sense here.

For the *final* evaluation of different approaches from participated teams, the task organizers allow three prediction results on the *final* test set. We used the prediction results given by `ensemble[4,5,7,8,9]`, `ensemble[4,5,7,8,9,10]`, `ensemble[4,5,6,7,8,9,10]` for the *final* test set provided by the task organizers, and our best-performing approach achieves the weighted-average F1 score of 0.92 on the *final* test set.

4   Conclusions

This paper shows how the proposed solution with different sentence encoders powered by transformers together with entity embeddings achieves the best classification performance in the context of a scholarly text classification task on the internal test set, which we believe can be generally applied to other text classification tasks as well. To the best of our knowledge, this is the first work using both sentence and entity embeddings for a text classification in the context of shared task or challenge, which might provide interesting insights for designing solutions for similar text classification tasks or challenges.

References

1. Beltagy, I., Lo, K., Cohan, A.: SciBERT: pretrained language model for scientific text. In: EMNLP (2019)
2. Cer, D., et al.: Universal sentence encoder. arXiv preprint arXiv:1803.11175 (2018)
3. Devlin, J., Chang, M.W., Lee, K., Toutanova, K.: BERT: pre-training of deep bidirectional transformers for language understanding. arXiv preprint arXiv:1810.04805 (2018)
4. Ferragina, P., Scaiella, U.: TAGME: on-the-fly annotation of short text fragments (by wikipedia entities). In: Proceedings of the 19th ACM International Conference on Information and Knowledge Management, pp. 1625–1628 (2010)
5. Ghosal, T., Sonam, R., Saha, S., Ekbal, A., Bhattacharyya, P.: Investigating domain features for scope detection and classification of scientific articles (2018)
6. Joulin, A., Grave, E., Bojanowski, P., Mikolov, T.: Bag of tricks for efficient text classification. arXiv preprint arXiv:1607.01759 (2016)
7. Piao, G., Breslin, J.G.: Domain-aware sentiment classification with GRUs and CNNs. In: Buscaldi, D., Gangemi, A., Reforgiato Recupero, D. (eds.) SemWebEval 2018. CCIS, vol. 927, pp. 129–139. Springer, Cham (2018). https://doi.org/10.1007/978-3-030-00072-1_11

8. Reddy, S., Saini, N.: Overview and insights from scope detection of the peer review articles shared tasks 2021. In: The First Workshop & Shared Task on Scope Detection of the Peer Review Articles (SDPRA 2021) (2018)

9. Reddy, S., Saini, N.: SDPRA 2021 shared task data (2021). Mendely Data, V1. https://doi.org/10.17632/njb74czv49.1

10. Reimers, N., Gurevych, I.: Sentence-BERT: sentence embeddings using Siamese BERT-networks. In: Proceedings of the 2019 Conference on Empirical Methods in Natural Language Processing. Association for Computational Linguistics (2019). https://arxiv.org/abs/1908.10084

11. Türker, R., Zhang, L., Alam, M., Sack, H.: Weakly supervised short text categorization using world knowledge. In: Pan, J.Z., et al. (eds.) ISWC 2020. LNCS, vol. 12506, pp. 584–600. Springer, Cham (2020). https://doi.org/10.1007/978-3-030-62419-4_33

12. Türker, R., Zhang, L., Koutraki, M., Sack, H.: "The less is more" for text classification, In: Semantics Posters& Demos (2018)

13. Vaswani, A., et al.: Attention is all you need. In: Advances in Neural Information Processing Systems, pp. 5998–6008 (2017)

14. Yamada, I., et al.: Wikipedia2Vec: an efficient toolkit for learning and visualizing the embeddings of words and entities from Wikipedia. In: Proceedings of the 2020 Conference on Empirical Methods in Natural Language Processing: System Demonstrations, pp. 23–30. Association for Computational Linguistics (2020)

Domain Identification of Scientific Articles Using Transfer Learning and Ensembles

Adeep Hande[1(✉)], Karthik Puranik[1], Ruba Priyadharshini[2],
and Bharathi Raja Chakravarthi[3]

[1] Indian Institute of Information Technology Tiruchirappalli,
Tiruchirappalli, Tamil Nadu, India
{adeeph18c,karthikp18c}@iiitt.ac.in
[2] ULTRA Arts and Science College, Madurai, Tamil Nadu, India
[3] Insight SFI Centre for Data Analytics, National University of Ireland Galway,
Galway, Ireland
bharathi.raja@insight-centre.org

Abstract. This paper describes our transfer learning-based approach for domain identification of scientific articles as a part of the SDPRA-2021 Shared Task. We experiment with transfer learning using pre-trained language models (BERT, RoBERTa, SciBERT), and these are then fine-tuned for this task. The result shows that the ensemble approach performs best as the weights are being taken into consideration. We propose improvements for future work. The codes for the best system are published here: https://github.com/SDPRA-2021/shared-task/tree/main/IIITT.

Keywords: Ensemble learning · Transfer learning · Sequence classification · Transformers

1 Introduction

Over the past few years, there has been an upswing in the number of scientific articles being published [1]. A study by bibliometric analysts demonstrate that the number of articles being published is said to be doubled in the next nine years, that is, 2021 [2]. This results in the need for automated recommendation systems to keep track of recent advancements in academic domains such as Computational Linguistics (CL), Cryptography and Security (CR), etc. The scientific articles are classified into relevant domains based on their abstracts. An abstract is a summary of a research article, thesis review, conference proceeding, or any in-depth analysis of a particular subject and is often used to help the reader quickly ascertain the paper's purpose[1].

---

[1] https://en.wikipedia.org/wiki/Abstract_(summary).

© Springer Nature Switzerland AG 2021
M. Gupta and G. Ramakrishnan (Eds.): PAKDD 2021 Workshops, LNAI 12705, pp. 88–97, 2021.
https://doi.org/10.1007/978-3-030-75015-2_9

This could be achieved by treating it as a sequence classification task in Natural Language Processing (NLP). Text classification is a common and important task of NLP to classify text and documents into predefined categories. As the number of articles build-up, retrieving papers of relevant academic categories will be cumbersome. This paper tries to address the above issue by classifying the abstract of scientific articles into predefined domains. The Shared task on Scope Detection of the Peer Review Articles for SDPRA-2021 emphasises on sorting scientific articles to store and retrieve a large number of research articles and in building personal recommendation systems [3]. The task of classifying the abstracts into seven domains of Computer Science was efficiently performed by Ensemble Classifier of Pre-trained Transformer-based models which are elaborated in this paper.

2 Related Work

Text classification in the scientific domain can be strenuous due to the absence of high-quality labelled scientific data in a large-scale. Subject text classification has been approached using deep neural networks with LSTM units on Wikipedia articles belonging to 7 subject categories [4] and by categorising them based on keywords [5]. Unigram and bigram are generated for these keywords on which query enrichment was used.

Subject classification was also tried using supervised learning by using links like citations, common authors, and common references to find an interrelationship between them [6]. The need for classifying the abstract sentences into its primary components in order to promote scientific database querying, summarise literature work, and the benefit in writing new abstracts was recognized and a deep learning classifier was trained on a repository containing 20 thousand abstracts from the biomedical domain to get impressive F1 scores [7]. There have also been works in the field of Big Data to retrieve information, extract document abstracts, or discover patterns to show promising results. Naïve Bayes classification algorithm served the following purposes on the Turkish scientific documents which were run on Cloud Computing infrastructure [8].

Ensemble transfer learning is observed to have boosted the classification accuracy when the dataset for training is limited [9]. There are studies on Bagging and Boosting [10], the two methods to produce ensemble which is more accurate than single classifiers [11]. Different ensemble approaches including ranking algorithms, polarity classifier, and a semisupervised system were used for Twitter sentiment analysis [12]. Works on predictive ensemble models using two BERT-based models like RoBERTa and ALBERT were seen to have outperformed the base model by an improvement of 3% in the F1 score [13].

3 Dataset

The dataset for this task is a collection of 35,000 abstracts of scientific articles spread over 7 predefined domains of Computer Science [14]. The distribution of the data is tabulated in Table 1.

1. **Computation and Language (CL)**
   It can be defined as an application of computer science to the analysis, synthesis and comprehension of written and spoken language.
2. **Cryptography and Security (CR)**
   It refers to the study of techniques for secure communication are included in this domain.
3. **Distributed and Cluster Computing (DS)**
   It refers to the application of cluster and distributed computing systems in order to leverage processing power, memory, etc., of any computing system for an increase in the efficiency.
4. **Data Structures and Algorithms (DS)**
   It refers to the study of using an entity that stores and organizes data in order to develop algorithms that help to decode and optimize computer programs.
5. **Logic in Computer Science (LO)**
   It refers to the overlap between the field of logic and that of computer science.
6. **Networking and Internet Architecture (NI)**
   It refers to the specification of a network's physical components and their functional organization and configuration.
7. **Software Engineering (SE)**
   It is the systematic application of engineering approaches to the development of software.

Table 1. Dataset Statistics

Category	Train	Validation	Test
CL	2,740	1,866	1,194
CR	2,660	1,835	1,105
DC	2,042	1,355	803
DS	2,737	1,774	1,089
LO	1,811	1,217	772
NI	2,764	1,826	1,210
SE	2,046	1,327	827
Total	**16,800**	**11,200**	**7,000**

4 System Description

4.1 System Architecture

In this section, we consider three kinds of approaches to classify the abstract. Figure 1 details the system architecture. We now describe how all the different modules are tied together. The input raw text is preprocessed as described in Sect. 4.2. The processed text is passed through all models described in Sects. 4.3, 4.4 and 4.5. Finally, the system returns the average of the predicted probabilities from all models as the output.

4.2   Data Preprocessing

As the text in the dataset in Table 1 primarily consists of abstracts of scientific articles, all words are lower-cased. We use the nltk library[2] to remove all expressions, equations and numbers, only to retain English words. For tokenization of the text, we use Huggingface's AutoTokenizer[3] to load the pretrained tokenizer in accordance with the needs.

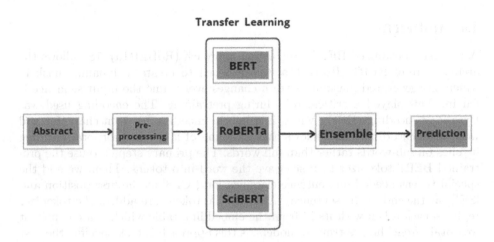

Fig. 1. System architecture

4.3   BERT

Bidirectional Encoder Representations from Transformers (BERT) [15] pretrained their language models on English Wikipedia (2.5B words) and Book-Corpus (0.8B words) [16]. The inputs are tokenized using BERT Tokenizer. It randomly masks 15% of the input tokens to train a deep bidirectional language model, and then to predict the masked token. This creates deep bidirectional representations by jointly conditioning on both left and right context in all layers. It is pretrained on 2 tasks, Next Sentence Prediction (NSP) and Masked Language Modelling (MLM).

The maximum length was set to 128, with shorter sequences padded and longer sequences truncated. Adam algorithm with weight decay fix algorithm [17] is used as the optimizer. We use the PyTorch implementation using the pre-trained BERT model **bert-base-uncased** from Huggingface[4]. The learning

---

[2] https://www.nltk.org/.

[3] https://huggingface.co/transformers/model_doc/auto.html#autotokenizer.

[4] https://huggingface.co/transformers/pretrained_models.html.

rate is set to $2e^{-5}$ with Cross-Entropy Loss as the loss function defined below:

$$L_{CE} = \sum_i^C t_i \log(s_i)$$

Where t_i and s_i are the ground truths for each class i in C. We use the same set of parameters for our other experiments.

4.4   RoBERTa

A Robustly optimized BERT pretraining approach (RoBERTa) [18] follows the architecture of BERT. RoBERTa was designed to create a dynamic mask in which the generated masking pattern changes every time the input sequence is fed in. This played a critical role during pretraining. The encoding used was Byte-Pair encoding (BPE), which is a hybrid encoding between character and word level encoding which allows easy handling of large text corpora, meaning it relies on sub-words rather than full words. The primary step is to use the pre-trained BERT tokenizer to first cleave the word into tokens. Then, we add the special tokens needed for sentence classification ([CLS] at the first position and [SEP] at the end of the sentence). After special tokens are added, the tokenizer replaces each token with its id from the embedding table which is a component we obtain from the pre-trained model. As this approach is task-specific, the first vector (the one associated with the [CLS] token) is passed on to a linear layer for classification.

4.5   SciBERT

We then tried using domain-specific variants of BERT and SciBERT [19]. SciB-ERT is a pretrained language model developed to address the lack of large anno-tated data in the scientific domain. This language model is based on BERT [15]. SciBERT was trained on 1.14M scientific papers, 15% in the domain of Computer Science and 85% in Biology. Sci-BERT consists of a custom-made vocabulary (Sci-Vocab), primarily consisting of frequently observed words and subwords in the scientific text which differ from those that may occur in the general domain text [20]. We used Allenai's[5] pretrained model *scibert_scivocab_uncased* hosted on huggingface's Model interface.

4.6   Ensembling

The prior probabilities of the predicted labels are computed as shown in Fig. 2. We combine the probabilities of all models above by the following two approaches:

---

[5] https://allenai.org/.

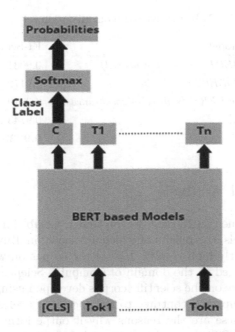

Fig. 2. Computing prior probabilities of the predicted labels

– **Average of probabilities** (E_1)
Taking the unweighted average of the posterior probabilities for these models, and the final prediction is the class with the largest averaged probability. We use the softmax function to calculate the posterior probabilities.
The softmax function is computed as follows:

$$e(z)_i = \frac{e^{z_i}}{\sum_{j=1}^{K} e^{z_j}}$$

– **Weighted Majority** (E_2)
We compute the Pearson correlation between the predicted probabilities to calculate the weights of each model [21]. Pearson correlation computes the linear relationship between 2 variables. This is a quantitative method to gauge model diversity. Weighted correlation is used to calculate the resultant probabilities. Then we compute the weighted average of the 3 models. One of the main reasons to use correlation methods is to calculate the weighted average of less correlated, but high performing models. The weights used are 0.3, 0.3 and 0.4 for BERT, RoBERTa, and SciBERT, respectively.
The Pearson's correlation coefficient (r) between 2 variables x and y is computed as follows:

$$r = \frac{\sum(x_i - \bar{x})(y_i - \bar{y})}{\sqrt{\sum(x_i - \bar{x})^2 \sum(y_i - \bar{y})^2}}$$

Table 2. F1-score on validation set

Model	F1-Score
BERT (bert-base-uncased)	0.9031
RoBERTa (roberta-base)	0.9172
SciBERT (scibert-scivocab-uncased)	0.9219
E_1	0.9143
E_2	**0.9229**

5   Results and Analysis

All models were trained on 16 GB GPU on Google Colab[6] for 10 epochs. We have
developed five models, the results of which are shown in Table 2. We see that all
models retrieved fairly high results. 15% of the corpus on which SciBERT was
pretrained on belonged to the domain of Computer Science. SciVocab is a new
wordpiece vocabulary on the scientific corpus developed using the SentencePiece
library. This was employed contrary to the vocabulary released with *BERT −
base*. We believe these are the reasons why it outperformed both BERT and
RoBERTa. As RoBERTa uses a dynamic masking pattern which masks the data
differently, it outperforms BERT.

Table 3. F1-score of system runs on test set

Model	F1-Score
SciBERT	0.9227
E_1	0.9156
E_2	**0.9246**

We experiment with an ensemble approach to check if there is any scope of
improvement in their F1 scores. Table 3 shows the system runs on the test set.
The classification report of the best system is shown in Table 4. We observe that
SciBERT performs better than the system $E1$. This is due to the correlation
between the models. As BERT, RoBERTa, and SciBERT are all highly reliant
on transformer blocks due to its success over recurrent neural networks [22], the
models are highly correlated. The approaches of averaging the probabilities tend
to work better if the models are less correlated with each other. However, the
system E_2 performs better than the two due to the weights being taken into
consideration. E_2 was the best system run on the test set.

---

[6] https://colab.research.google.com/.

Table 4. Classification report of the best system(E_2)

Domain	Precision	Recall	F1-score	Support
DS	0.99	0.98	0.98	1194
NI	0.91	0.93	0.92	1105
CR	0.84	0.83	0.84	803
LO	0.94	0.93	0.93	1089
DC	0.94	0.93	0.94	772
SE	0.93	0.92	0.92	1210
CL	0.90	0.93	0.91	827
Accuracy			0.92	7000
Macro avg	0.92	0.92	0.92	7000
Weighted avg	0.92	0.92	0.92	7000

6   Conclusion

In this paper, we describe our methods for domain identification of scientific articles. We have presented three models out of which we achieved the best result using an ensemble of transfer learning models. We demonstrate that ensembles of transfer learning using pre-trained language models yield the highest performance. Our work was highly inspired by the exceptional performance of ensemble models in Kaggle competitions[7]. The better performance is due to the diversification or independent nature of the independent models. We explore the collinearity between the base models in order to compute the weights of each base model.

References

1. Cohan, A., Goharian, N.: Scientific article summarization using citation-context and article's discourse structure. In: Proceedings of the 2015 Conference on Empirical Methods in Natural Language Processing, pp. 390–400. Association for Computational Linguistics, Lisbon, September 2015. https://doi.org/10.18653/v1/D15-1045. https://www.aclweb.org/anthology/D15-1045
2. Bornmann, L., Mutz, R.: Growth rates of modern science: a bibliometric analysis based on the number of publications and cited references. J. Assoc. Inf. Sci. Technol. **66**(11), 2215–2222 (2015). https://doi.org/10.1002/asi.23329. https://asistdl.onlinelibrary.wiley.com/doi/abs/10.1002/asi.23329
3. Reddy, S., Saini., N.: Overview and insights from scope detection of the peer review articles shared tasks 2021. In: In: Proceedings of the The First Workshop & Shared Task on Scope Detection of the Peer Review Articles (SDPRA 2021) (forthcoming)
4. Semberecki, P., Maciejewski, H.: Deep learning methods for subject text classification of articles, pp. 357–360 (2017). https://doi.org/10.15439/2017F414

---

[7] https://www.kaggle.com/c/jigsaw-unintended-bias-in-toxicity-classification/discussion/103280.

5. Roul, R., Sahoo, J.: Classification of research articles hierarchically: a new technique, pp. 347–361, May 2017. https://doi.org/10.1007/978-981-10-3874-7_32
6. Taheriyan, M.: Subject classification of research papers based on interrelationships analysis. In: Proceedings of the 2011 Workshop on Knowledge Discovery, Modeling and Simulation, KDMS 2011, pp. 39–44. Association for Computing Machinery, New York (2011). https://doi.org/10.1145/2023568.2023579
7. Gonçalves, S., Cortez, P., Moro, S.: A deep learning classifier for sentence classification in biomedical and computer science abstracts. Neural Comput. Appl. **32**(11), 6793–6807 (2019). https://doi.org/10.1007/s00521-019-04334-2
8. Gurbuz, S., Aydin, G.: Classification of scientific papers with big data technologies. In: 2017 International Conference on Computer Science and Engineering (UBMK), pp. 697–701 (2017). https://doi.org/10.1109/UBMK.2017.8093504
9. Liu, X., Liu, Z., Wang, G., Cai, Z., Zhang, H.: Ensemble transfer learning algorithm. IEEE Access **6**, 2389–2396 (2018). https://doi.org/10.1109/ACCESS.2017.2782884
10. Bühlmann, P.: Bagging, boosting and ensemble methods. In: Härdle, W., Mori, Y. (eds.) Handbook of Computational Statistics, pp. 985–1022. Springer, Heidelberg (2012). https://doi.org/10.1007/978-3-642-21551-3_33
11. Opitz, D., Maclin, R.: Popular ensemble methods: an empirical study. J. Artif. Int. Res. **11**(1), 169–198 (1999)
12. Ankit, Saleena, N.: An ensemble classification system for twitter sentiment analysis. Proc. Comput. Sci. **132**, 937–946 (2018). https://doi.org/10.1016/j.procs.2018.05.109. http://www.sciencedirect.com/science/article/pii/S187705091830841X. International Conference on Computational Intelligence and Data Science
13. Dadu, T., Pant, K., Mamidi, R.: BERT-based ensembles for modeling disclosure and support in conversational social media text, June 2020
14. Reddy, S., Saini, N.: SDPRA 2021 shared task data (2021). https://doi.org/10.17632/NJB74CZV49.1. https://data.mendeley.com/datasets/njb74czv49/1
15. Devlin, J., Chang, M.W., Lee, K., Toutanova, K.: BERT: pre-training of deep bidirectional transformers for language understanding. In: Proceedings of the 2019 Conference of the North American Chapter of the Association for Computational Linguistics: Human Language Technologies, Volume 1 (Long and Short Papers), pp. 4171–4186. Association for Computational Linguistics, Minneapolis, June 2019. https://doi.org/10.18653/v1/N19-1423. https://www.aclweb.org/anthology/N19-1423
16. Zhu, Y., et al.: Aligning books and movies: towards story-like visual explanations by watching movies and reading books. CoRR abs/1506.06724 (2015). http://arxiv.org/abs/1506.06724
17. Loshchilov, I., Hutter, F.: Fixing weight decay regularization in Adam (2018). https://openreview.net/forum?id=rk6qdGgCZ
18. Liu, Y., et al.: Roberta: a robustly optimized BERT pretraining approach (2020). https://openreview.net/forum?id=SyxS0T4tvS
19. Beltagy, I., Lo, K., Cohan, A.: SciBERT: a pretrained language model for scientific text. In: Proceedings of the 2019 Conference on Empirical Methods in Natural Language Processing and the 9th International Joint Conference on Natural Language Processing (EMNLP-IJCNLP), pp. 3615–3620. Association for Computational Linguistics, Hong Kong, November 2019. https://doi.org/10.18653/v1/D19-1371. https://www.aclweb.org/anthology/D19-1371
20. Sharma, P., Roychowdhury, S.: IIT-KGP at MEDIQA 2019: recognizing question entailment using sci-BERT stacked with a gradient boosting classifier. In: Proceedings of the 18th BioNLP Workshop and Shared Task, pp. 471–477. Association for

Computational Linguistics, Florence, August 2019. https://doi.org/10.18653/v1/ W19-5050. https://www.aclweb.org/anthology/W19-5050

21. Kirch, W. (ed.): Pearson's Correlation Coefficient, pp. 1090–1091. Springer, Dordrecht (2008). https://doi.org/10.1007/978-1-4020-5614-7_2569

22. Vaswani, A., et al.: Attention is all you need. In: Guyon, I., Luxburg, U.V., Bengio, S., Wallach, H., Fergus, R., Vishwanathan, S., Garnett, R. (eds.) Advances in Neural Information Processing Systems, vol. 30, pp. 5998–6008. Curran Associates, Inc. (2017). https://proceedings.neurips.cc/paper/2017/file/ 3f5ee243547dee91fbd053c1c4a845aa-Paper.pdf

Identifying Topics of Scientific Articles with BERT-Based Approaches and Topic Modeling

Anna Glazkova$^{(\boxtimes)}$ ⓘ

University of Tyumen, ul. Volodarskogo 6, 625003 Tyumen, Russia
`a.v.glazkova@utmn.ru`

Abstract. This paper describes neural models developed for the First Workshop on Scope Detection of the Peer Review Articles shared task collocated with PAKDD 2021. The aim of the task is to identify topics or category of scientific abstracts. We investigate the use of several fine-tuned language representation models pretrained on different large-scale corpora. In addition, we conduct experiments on combining BERT-based models and document topic vectors for scientific text classification. The topic vectors are obtained using LDA topic modeling. The topic-informed soft voting ensemble of neural networks achieved F1-score of 93.82%.

Keywords: BERT · Ensembling learning · Topic modeling · Scientific text · Scholarly documents · SciBERT · LDA

1 Introduction

Electronic scientific archives and other scholarly document repositories are sources of a large number of significant facts and valuable research information. The electonic sources are rapidly expanding. They contain various and mul-tithematic scientific work. Therefore, the systematization of stored documents is an important task of natural language processing (NLP) and informational retrieval.

This paper describes our system presented at the First Workshop & Shared task on Scope Detection of the Peer Review Articles collocated with PAKDD 2021. The shared task aimed at identifying topical categories of scientific articles. Our system is an ensemble of transformer-based and topic-informed neural models. We use both SciBERT language model and Latent Dirichlet allocation (LDA) topic modeling to increase the quality of classification. During the final stage of the competition, our model achieved 93.82% of the weighted F1-score.

The paper is organized as follows. A brief review of related work is given in Sect. 2. Section 3 provides the main details related to the shared task definition and the dataset. Section 4 contains the description of the models used. In Sect. 5, we describe our experiments and results. Section 5 is a conclusion.

© Springer Nature Switzerland AG 2021
M. Gupta and G. Ramakrishnan (Eds.): PAKDD 2021 Workshops, LNAI 12705, pp. 98–105, 2021.
https://doi.org/10.1007/978-3-030-75015-2_10

2   Related Work

Electronic archives have considerable potential and wide opportunities for the development of NLP methods in the field of scientific text classification [6,29], analysis of their structure [12,32], information extraction [3], and other applications of computational linguistics for the tasks related to science and education [41]. Thus, the methods of NLP can both help to assign a new scientific text to one of the specified categories, and organize a search in a set of existing documents.

There was also a related previous work on collection of scientific text corpora for text classification [19,25,31,34,36], summarization [9,30,40], concept understanding [11], and citation analysis [8,13]. The organization of competitions and the publication of shared tasks significantly contribute to the growth of interest in the scientific field. Thus, in recent years, several competitions [4,10,35,37] related to the analysis of scientific texts have been held.

3   Shared Task

3.1   Task Definition

The task focused on identifying topics or category of scientific articles. The sources of data was 35000 abstracts scientific article (computer science) from different corpora. Formally, the task is described as follows.

- **Input.** Given an abstract.
- **Output.** One of 7 predefined domains.

The official evaluation metric was the weighted-average F1-score. The participants were allowed three submissions per team throughout the test phase.

3.2   Dataset

We use the officially provided dataset [27]. This dataset includes abstracts of scientific articles in 7 research areas. The distribution of the data is shown in Table 1.

As can be seen from the table above, the number of annotations in categories is not the same. The "Sum" column shows the total number of texts for the research area. It can be seen from the data in Table 1 that Networking and Internet Architecture (2764 texts in the training set and 5800 texts in total) and Computation and Language (2740 texts in the training set and 5800 texts in total) are majority classes. The minority class is Logic in Computer Science (1811 texts in the training set and 3800 texts in total). Thus, the number of texts in the largest category are more than one-and-one-half times greater than the number of texts in the smallest category.

Table 1. Statistics of the dataset.

Category	Train	Validation	Test	Sum
Computation and Language (CL)	2740	1866	1194	5800
Cryptography and Security (CR)	2660	1835	1105	5600
Distributed and Cluster Computing (DC)	2042	1355	803	4200
Data Structures and Algorithms (DS)	2737	1774	1089	5600
Logic in Computer Science (LO)	1811	1217	772	3800
Networking and Internet Architecture (NI)	2764	1826	1210	5800
Software Engineering (SE)	2046	1327	827	4200
Total	16800	11200	7000	35000

4   Methodology

We compared several scope detection models. All these models are based on the transformer architecture which is state-of-the art for many NLP tasks. In addition, language models trained on specific domain datasets are often superior in quality to models based only on general corpora for various narrowly defined tasks (in particular, it was shown in [1,15,22,33], see also a survey [39]). Below is a list of the evaluated models.

- **BERT** [7]. BERT is a language representation model presented by Google, which stands for Bidirectional Encoder Representations from Transformers. BERT-based models show great results in many natural language processing tasks. In our work, we used BERT-base-uncased, which is pretrained on texts from Wikipedia.
- **RoBERTa** [18]. RoBERTa is a robustly optimized BERT approach introduced at Facebook. Unlike BERT, RoBERTa removes the Next Sentence Prediction task from the pretraining process. RoBERTa also uses larger batch sizes and dynamic masking so that the masked token changes while training instead of the static masking pattern used in BERT. We experimented with RoBERTa-large.
- **SciBERT** [2]. SciBERT is a BERT-based model trained on scientific texts. The training corpus was a set of papers taken from Semantic Scholar. The authors used the full text of the papers in training, not only abstracts. The size of the training corpus was 1.14M papers, 3.1B tokens. SciBERT also has its own wordpiece vocabulary (scivocab) that was built to best match the training corpus. In our experiments, we used SciBERT-scivocab-uncased.
- **PubMedBERT** [16]. This model is pretrained from scratch using abstracts from PubMed. PubMedBERT achieves state-of-the-art performance on several biomedical NLP tasks related to biomedical language understanding. The training corpus for this model is out of the computer science domain, but we evaluated this model because it was trained on scientific abstracts. In our work, we experimented with BiomedNLP-PubMedBERT-base-uncased-abstract.

– **Topic-informed BERT-based model.** This model combines forces of BERT-based transformers and topic modeling approaches. We were inspired by [24] where a topic-informed BERT-based architecture (tBERT) was applied to the task of semantic similarity detection. We made an attempt to adapt this approach for scope detection in the scientific domain. The architecture we used is shown in Fig. 1.

As can be seen from the figure, the concatenation of the text embedding derived from the BERT-based model and the document topic distribution obtained with topic modeling is an input layer for the topic-informed feedforward neural network (FNN).

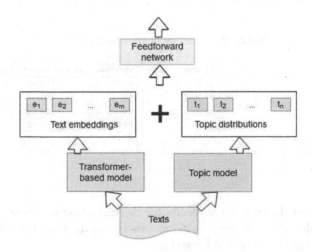

Fig. 1. The architecture of the topic-informed BERT-based model.

5 Experiments and Results

We conducted our experiments on Google Colab Pro (CPU: Intel(R) Xeon(R) CPU @ 2.20 GHz; RAM: 25.51 GB; GPU: Tesla P100-PCIE-16 GB with CUDA 10.1). Each language model was fine-tuned on the training set for 3 epochs and evaluated on the validation set. The models were optimised using AdamW [21] with a learning rate of 2e−5 and epsilon of 1e−8, max sequence length of 256 tokens, and a batch size of 8. We implemented our models using Pytorch [23] and Huggingface's Transformers [38].

For the topic-informed model, we built an LDA topic model for 100 topics. The main preprocessing steps were translating to lower case, removing punctuation and digits, lemmatizing, removing extra white spaces, excluding stop words, and combining common phrases into bigrams. We also used a random oversampling technique that is able to effectively and low-resource work with class imbalance [14]. Random oversampling was applied to all categories except the majority category in the training set(Networking and Internet Architecture, 2764 texts). The texts of the minority categories were sampled until the number of examples in all categories was equal. Thus, the final size of the training sample was 19343 (2764 × 7).

We used Gensim [28] and NLTK [20] Python libraries for text preprocessing and topic modeling. The random sampling of minority class examples was implemented using Imbalanced-learn [17]. The FNN was implemented using Keras [5]. It contained two fully-connected hidden layers with 2048 and 1024 neurons, respectively. The activation function for hidden layers was hyperbolic tangent and for the output layer it was softmax. The cost function was categorical crossentropy. We trained the FNN for 5 epoches.

Table 2. Model evaluation results (%).

Model	F1-score (macro)	Precision (macro)	Recall (macro)	Accuracy
BERT	91.4	91.47	91.35	91.76
RoBERTa	92.03	92.07	92.01	92.36
SciBERT	92.91	92.95	92.88	93.19
PubMedBERT	91.75	91.79	91.72	92.08
FNN (SciBERT embeddings)	92.79	92.75	92.88	93.11
FNN (SciBERT embs + document topics)	92.99	93	92.99	93.28

Table 2 shows the evaluation results for the validation set. Note that we evaluated our models using F1-score with macro-averaging while the official competition metric was the weighted F1-score.

As can be seen from the table above, SciBERT showed absolutely better results compared to other fine-tuning BERT-based models. However, the best results on the validation set were shown by the topic-informed FNN model trained on SciBERT embeddings and document topic vectors. It exceeded the F1-score of SciBERT by 0.08% and the accuracy by 0.09%. For comparison, we showed the result of the feedforward model trained only on SciBERT embeddings, which is lower than the results of the fine-tuning SciBERT and the topic-informed FNN.

For final submissions[1], we chose SciBERT and the topic-informed FNN because these models showed the closest and highest results on the validation set. Each model was trained on both original training and validation sets randomly splitted into train and validation subsets in the ratio of 90:10. Next, we used a soft voting ensemble of three SciBERT models with random seed values (Table 3, submission 1), a soft voting ensemble of three topic-informed FNNs (submission 2), and a soft voting ensemble of the first two models (submission 3).

---

[1] The source code for our models is available at: https://github.com/SDPRA-2021/shared-task/tree/main/utmn

Table 3. Final results (%).

Submission ID	Model	F1-score (weighted)
1	SciBERT ensemble	93.7
2	FNN (SciBERT embs + document topics) ensemble	93.54
3	SciBERT and FNN ensemble	93.82

The best result (93.82% of weighted F1-score) was achieved by the third submitted model. In contrast to the validation set, topic-informed models performed worse than SciBERT on the test set.

6   Conclusion

In this work, we propose an approach to identifying topic of scientific articles based on SciBERT, topic modeling, and ensembling learning. Our experiments confirmed that adding topic modeling features can improve the quality of topical text classification in the scientific domain. In addition, the topic-informed BERT-based model performed better than pretrained SciBERT. However, the SciBERT ensemble showed the higher result than the topic-informed FNN ensemble for the test set.

The experimental results showed that our solution achieved 93.82% of the weighted F1-score on test data. For future work, we can experiment with different topic models and datasets as well as various ensemble architectures.

References

1. Aluru, S.S., et al.: Deep learning models for multilingual hate speech detection. arXiv preprint arXiv:2004.06465 (2020)
2. Beltagy, I., Lo, K., Cohan, A.: SciBERT: a pretrained language model for scientific text. arXiv preprint arXiv:1903.10676 (2019)
3. Bidulya, Y.: An Approach to the development of software for effective search of scientific articles. In: 2018 3rd Russian-Pacific Conference on Computer Technology and Applications (RPC), pp. 1–4 (2018). https://doi.org/10.1109/rpc.2018.8482164
4. Chandrasekaran, M.K., et al.: Overview and insights from the shared tasks at scholarly document processing 2020: CL-SciSumm, LaySumm and LongSumm. In: Proceedings of the First Workshop on Scholarly Document Processing, pp. 214–224 (2020)
5. Chollet, F., et al.: Keras: the python deep learning library. Astrophysics Source Code Library, ascl: 1806.022 (2018)
6. Cox, J., Harper, C.A., de Waard, A.: Optimized machine learning methods predict discourse segment type in biological research articles. In: González-Beltrán, A., Osborne, F., Peroni, S., Vahdati, S. (eds.) SAVE-SD 2017-2018. LNCS, vol. 10959, pp. 95–109. Springer, Cham (2018). https://doi.org/10.1007/978-3-030-01379-0_7
7. Devlin, J., et al.: BERT: pre-training of deep bidirectional transformers for language understanding. arXiv preprint arXiv:1810.04805 (2018)

8. Fisas, B., Ronzano, F., Saggion, H.: A multi-layered annotated corpus of scientific papers. In: Proceedings of the Tenth International Conference on Language Resources and Evaluation (LREC 2016), pp. 3081–3088 (2016)
9. Jaidka, K., et al.: Insights from CL-SciSumm 2016: the faceted scientific document summarization shared task. Int. J. Digit. Libr. **192**, 163–171 (2018)
10. Gábor, K., et al.: Semeval-2018 task 7: semantic relation extraction and classification in scientific papers. In: Proceedings of The 12th International Workshop on Semantic Evaluation, pp. 679–688 (2018)
11. Gordon, J., et al.: Modeling concept dependencies in a scientific corpus. In: Proceedings of the 54th Annual Meeting of the Association for Computational Linguistics (Volume 1: Long Papers), pp. 866–875 (2016)
12. Ghosal, T., Verma, R., Ekbal, A., Saha, S., Bhattacharyya, P.: An empirical study of importance of different sections in research articles towards ascertaining their appropriateness to a Journal. In: Ishita, E., Pang, N.L.S., Zhou, L. (eds.) ICADL 2020. LNCS, vol. 12504, pp. 407–415. Springer, Cham (2020). https://doi.org/10.1007/978-3-030-64452-9_38
13. Gipp, B., Meuschke, N., Breitinger, C.: Citation based plagiarism detection: practicability on a large scale scientific corpus. J. Am. Soc. Inf. Sci. **658**, 1527–1540 (2014). https://doi.org/10.1002/asi.23228
14. Glazkova, A.: A comparison of synthetic oversampling methods for multi-class text classification. arXiv preprint arXiv:2008.04636 (2020)
15. Glazkova, A., Glazkov, M., Trifonov, T.: g2tmn at Constraint@AAAI2021: exploiting CT-BERT and ensembling learning for COVID-19 fake news detection. In: Proceedings of the First Workshop on Combating Online Hostile Posts in Regional Languages during Emergency Situation (CONSTRAINT). Springer (2021, Forthcoming)
16. Gu, Y., et al.: Domain-specific language model pretraining for biomedical natural language processing. arXiv preprint arXiv:2007.15779 (2020)
17. Lemaître, G., Nogueira, F., Aridas, C.K.: Imbalanced-learn: a python toolbox to tackle the curse of imbalanced datasets in machine learning. J. Mach. Learn. Res. **181**, 559–563 (2017)
18. Liu, Y., et al.: Roberta: a robustly optimized bert pretraining approach. arXiv preprint arXiv:1907.11692 (2019)
19. Lo, K., et al.: S2ORC: the semantic scholar open research corpus. In: Proceedings of the 58th Annual Meeting of the Association for Computational Linguistics, pp. 4969–4983 (2020)
20. Loper, E., Bird, S.: NLTK: the natural language toolkit. In: Proceedings of the ACL-02 Workshop on Effective Tools and Methodologies for Teaching Natural Language Processing and Computational Linguistics, pp. 63–70 (2002)
21. Loshchilov I., Hutter F.: Fixing weight decay regularization in Adam. arXiv preprint arXiv:1711.05101 (2017)
22. Müller, M., Salathé, M., Kummervold, P.E.: Covid-Twitter-BERT: a natural language processing model to analyse covid-19 content on twitter. arXiv preprint arXiv:2005.07503 (2020)
23. Paszke, A., et al.: PyTorch: an imperative style, high-performance deep learning library. In: Advances in Neural Information Processing Systems, pp. 8026–8037 (2019)
24. Peinelt N., Nguyen D., Liakata M.: tBERT: Topic models and BERT joining forces for semantic similarity detection. In: Proceedings of the 58th Annual Meeting of the Association for Computational Linguistics, pp. 7047–7055 (2020) https://doi.org/10.18653/v1/2020.acl-main.630

25. Radev, D.R., et al.: The ACL anthology network corpus. Lang. Resour. Eval. **474**, 919–944 (2013). https://doi.org/10.3115/1699750.1699759
26. Reddy, S., Saini, N.: Overview and insights from scope detection of the peer review articles shared tasks 2021. In: Proceedings of The First Workshop & Shared Task on Scope Detection of the Peer Review Articles (SDPRA 2021) (2021, Forthcoming)
27. Reddy S., Saini N.: SDPRA 2021 shared task data. Mendeley Data V1. https://doi.org/10.17632/njb74czv49.1
28. Rehurek, R., Sojka, P.: Software framework for topic modelling with large corpora. In: Proceedings of the LREC 2010 Workshop on New Challenges for NLP Frameworks (2010)
29. Romanov, A., Lomotin, K., Kozlova, E.: Application of natural language processing algorithms to the task of automatic classification of russian scientific texts. Data Sci. J. **181** (2019). https://doi.org/10.5334/dsj-2019-037
30. Saggion, H., et al.: A multi-level annotated corpus of scientific papers for scientific document summarization and cross-document relation discovery. In: Proceedings of The 12th Language Resources and Evaluation Conference, pp. 6672–6679 (2020)
31. Soares F., Moreira V., Becker K.: A large parallel corpus of full-text scientific articles. In: Proceedings of the Eleventh International Conference on Language Resources and Evaluation (LREC 2018) (2018)
32. Solovyev, V., Ivanov, V., Solnyshkina, M.: Assessment of reading difficulty levels in Russian academic texts: approaches and metrics. J. Intell. Fuzzy Syst. **345**, 3049–3058 (2018). https://doi.org/10.3233/jifs-169489
33. Sun, Z., et al.: MobileBERT: a compact task-agnostic BERT for resource-limited devices. In: Proceedings of the 58th Annual Meeting of the Association for Computational Linguistics, pp. 2158–2170 (2020)
34. Teich, E.: Exploring a corpus of scientific texts using data mining. In: Corpus-Linguistic Applications, pp. 233–247 (2010)
35. Veyseh, A.P.B., et al.: Acronym identification and disambiguation shared tasks for scientific document understanding. arXiv preprint arXiv:2012.11760 (2020)
36. Vincze, V., et al.: The BioScope corpus: biomedical texts annotated for uncertainty, negation and their scopes. BMC Bioinf. **911**, 1–9 (2008)
37. Weissenbacher, D., et al.: SemEval-2019 task 12: toponym resolution in scientific papers. In: Proceedings of the 13th International Workshop on Semantic Evaluation, pp. 907–916 (2019)
38. Wolf, T., et al.: Transformers: state-of-the-art natural language processing. In: Proceedings of the 2020 Conference on Empirical Methods in Natural Language Processing: System Demonstrations, pp. 38–45 (2020) https://doi.org/10.18653/v1/2020.emnlp-demos.6
39. Xia, P., Wu, S., Van Durme, B.: Which* BERT? A survey organizing contextualized encoders. In: Proceedings of the 2020 Conference on Empirical Methods in Natural Language Processing (EMNLP), pp. 7516–7533 (2020)
40. Yasunaga, M., et al.: ScisummNet: a large annotated corpus and content-impact models for scientific paper summarization with citation networks. In: Proceedings of the AAAI Conference on Artificial Intelligence, vol. 331, pp. 7386–7393 (2019)
41. Zakharova, I., et al.: Diagnostics of professional competence of IT students based on digital footprint data. Inf. Educ. **4**, 4–11 (2020). https://doi.org/10.32517/0234-0453-2020-35-4-4-11

Using Transformer Based Ensemble Learning to Classify Scientific Articles

Sohom Ghosh[(✉)] and Ankush Chopra

Artificial Intelligence, CoE, Fidelity Investments, Bengaluru, Karnataka, India

Abstract. Many time reviewers fail to appreciate novel ideas of a researcher and provide generic feedback. Thus, proper assignment of reviewers based on their area of expertise is necessary. Moreover, reading each and every paper from end-to-end for assigning it to a reviewer is a tedious task. In this paper, we describe a system which our team FideLIPI submitted in the shared task of SDPRA-2021 (https://sdpra-2021.github.io/website/ (accessed January 25, 2021)) [14]. It comprises four independent sub-systems capable of classifying abstracts of scientific literature to one of the given seven classes. The first one is a RoBERTa [10] based model built over these abstracts. Adding topic models/Latent dirichlet allocation (LDA) [2] based features to the first model results in the second sub-system. The third one is a sentence level RoBERTa [10] model. The fourth one is a Logistic Regression model built using Term Frequency Inverse Document Frequency (TF-IDF) features. We ensemble predictions of these four sub-systems using majority voting to develop the final system which gives a F1 score of 0.93 on the test and validation set. This outperforms the existing State Of The Art (SOTA) model SciBERT's [1] in terms of F1 score on the validation set. Our codebase is available at https://github.com/SDPRA-2021/shared-task/tree/main/FideLIPI.

Keywords: Scientific text classification · Ensemble learning · Transformers

1 Introduction

Due to the ever-increasing number of research paper submissions per conference, it has become extremely difficult to manually assign appropriate reviewers to a paper based on their expertise. Improper assignment of reviewer leads to poor quality of reviews. Thus, an automated system capable of determining which category a research paper belongs to is necessary to develop. A similar shared task has been hosted at The First Workshop & Shared Task on Scope Detection

---

[1] https://www.pakdd2021.org/ (accessed January 25, 2021).

S. Ghosh and A. Chopra—Equal Contribution.

© Springer Nature Switzerland AG 2021
M. Gupta and G. Ramakrishnan (Eds.): PAKDD 2021 Workshops, LNAI 12705, pp. 106–113, 2021.
https://doi.org/10.1007/978-3-030-75015-2_11

of the Peer Review Articles (SDPRA-2021) (Collocated with the 25th Pacific-Asia Conference on Knowledge Discovery and Data Mining[1] (PAKDD-2021)) [14]. This paper narrates the approach our team FideLIPI followed while participating in this challenge. It is structured into five main sections. This section introduces readers to the problem that we have tried to solve, related works that have been done previously and our contributions. The next section familiarizes readers with the dataset, pre-processing and feature engineering steps. The section following it describes the system we developed and how we generated the submissions we have made. The subsequent section discusses the experiments we performed and their results. After that, we conclude the paper by revealing our plans to improve the system further.

1.1 Problem Description

Given an abstract A of a scientific article, we need to assign one of the seven classes (CL, CR, DC, DS, LO, NI, SE)[2] to it.

1.2 Related Works

Beltagy et al. in their paper [1] propose SciBERT which is a language model trained specifically for scientific articles. They outperformed the BERT [7] base model in domains like Bio-medicine and Computer Science. In the paper [4] Cao et al. describe how contents as well as citations can be used to classify scientific documents. They try out different Machine Learning models like K Nearest Neighbours, Nearest Centroid and Naive Bayes. One of the pioneering work has been done by Ghanem et al. [8]. While participating in KDD CUP 2002 (Task 1) they proposed a novel method for extracting frequently appearing keywords. They use these patterns as input to a Support Vector Machine (SVM) classifier [6]. Borrajo et al. [3] studied how over-sampling, under-sampling, use of dictionaries help in classifying scientific articles relating to bio-medicine. They mainly dealt with three kinds of classifiers: K Nearest Neighbour, SVM and Naive-Bayes. For evaluation, they used Precision, Recall, F-measure and Utility. They achieved best results by using sub-sampling along with NLPBA, Protein dictionaries and a SVM classifier [6].

1.3 Our Contributions

Our contributions are as follows:

- We have developed a system capable of assigning a class to the abstract of a given scientific article. Our model (F1: 0.928) surpassed the existing SOTA model SciBERT's [1] performance (F1: 0.926) in the given validation set.
- For enhancing reproducibility and transparency, we have open-sourced the system. It is available here[3].

---

[2] These abbreviations have been expanded in Table 1.
[3] https://github.com/SDPRA-2021/shared-task/tree/main/FideLIPI.

2   Dataset

This section narrates in details the data we are dealing with and the pre-processing steps we followed.

2.1   Data Description

The dataset [13] for the shared task of The First Workshop and Shared Task on Scope Detection of the Peer Review Articles (SPDRA 2021) [14] consists of 16,800 training instances, 11,200 validation instances and 7,000 test instances. They belong to seven classes as described in Table 1.

Table 1. Data Description

Category	Train	Validation	Test
Computation and Language (CL)	2,740	1,866	1,194
Cryptography and Security (CR)	2,660	1,835	1,105
Distributed and Cluster Computing (DC)	2,042	1,355	803
Data Structures and Algorithms (DS)	2,737	1,774	1,089
Logic in Computer Science (LO)	1,811	1,217	772
Networking and Internet Architecture (NI)	2,764	1,826	1,210
Software Engineering (SE)	2,046	1,327	827
Total	**16,800**	**11,200**	**7,000**

2.2   Pre-processing and Feature Engineering

For Model-1 narrated in Sect. 3.1, we keep the raw abstracts as it is and we do not pre-process them. For Model-2, after converting the abstracts to lowercase, removing stop words and lemmatizing them, we empirically decide to extract 50 topics using Topic Modelling (LDA) [2]. This is narrated in Sect. 3.2. We used NLTK[4] and Gensim[5] libraries to achieve this. As pre-processing steps of Model-3 described in Sect. 3.3, we remove newline characters and extract sentences from abstracts of scientific articles using Spacy[6] library. For Model-4 mentioned in Sect. 3.4, we remove stop words and create TF-IDF based features with n-grams ranging from 1 to 4. We further ignore the terms with document frequency strictly lesser than 0.0005. We selected these hyper-parameters through experimentation. This resulted in 22,151 features.

---

[4] https://www.nltk.org/ (accessed January 25, 2021).
[5] https://radimrehurek.com/gensim/ (accessed January 25, 2021).
[6] https://spacy.io/ (accessed January 25, 2021).

3 Methodology

In this section, we describe each of the sub-systems/models which are ensembled to create the final system. Moreover, we elucidate the process we followed to generate each of the three submissions we have made.

NOTE: All of the hyper-parameters mentioned in this section have been obtained through rigorous experimentation described in Sect. 4.1. The hyper-parameters of the RoBERTa [10] models used here are mentioned in Table 3.

3.1 System Description of Model-1 (RoBERTa)

It is a RoBERTa [10] based model built on the raw text corpus of the abstract. Its task is to predict probability of each of the given classes. We finally select the class having maximum probability.

3.2 System Description of Model-2 (RoBERTa+LDA)

This model is similar to the one described in Sect. 3.1. It additionally takes 50 Topic Modelling (LDA) [2] based features. Additional dropout of 0.3 is implemented in the classification layer of this RoBERTa [10] model.

3.3 System Description of Model-3 (RoBERTa on Sentences)

In this model, instead of considering the whole abstract as input, we split it into individual sentences. Sentences having a number of tokens greater or equal to ten are considered while training the RoBERTa [10] based model. While scoring the validation set we considered only those sentences which have a number of tokens greater or equal to six. These numbers have been obtained empirically. To decide the final label of a given abstract, we add the logarithmic probabilities of predictions of the individual sentences for each of the seven classes. We then, select the class for which this value is maximum.

3.4 System Description of Model-4 (TF-IDF + Logistic Regression)

This is a simple logistic regression model built using scikit-learn[7] library over 22,151 TF-IDF features. Its hyper-parameters are as follows: maximum number of iterations = 100, penalty = l2, tolerance = 0.0001.

3.5 Submissions

As per the rules of this shared task, each team could submit at-most three sets of predictions for the test set. Our first submission is an ensemble of all the four models described above using majority voting technique. It has been depicted in Fig. 1. In this technique, the class which gets the maximum number of vote is selected as the final class. Whenever there is a tie between two classes, we randomly choose one of them. The second and third submissions are results of Model-2 (refer to Sect. 3.2) and Model-4 (refer to Sect. 3.4) respectively.

---

[7] https://scikit-learn.org/ (accessed January 25, 2021).

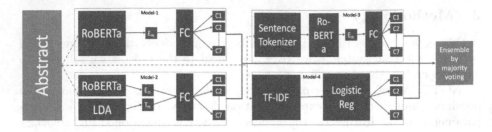

Fig. 1. Ensemble Model Architecture. E_m = Embedding, T_m = Topic Models, FC = Fully Connected Layer, C_1 to C_2 are the classes corresponding to CL, CR and so on

4   Experiments, Results and Discussion

4.1   Experimental Setup

We started with zero-shot learning [5] and pre-trained models like BERT [7], RoBERTa [10] and T5 [12], since the transformer-based pre-trained models are producing state-of-the-art results on most of the Natural Language Processing (NLP) tasks. We used text classification module of the Simple Transformer library[8] to run the multi-class classification using the BERT [7] model. The Simple Transformer library provides an easy interface to run different NLP tasks while using the HuggingFace [15] Transformers at the back-end. We fine-tune the underlying pre-trained model while training it for the task. We ran classification using BERT-base [7] for 10 epochs and saved model after each epoch. Based on the saturation of validation set performance improvement, we chose the right epoch for both the models.

T5 [12] uses both encoder and decoder parts of the transformer. Although both input and output of the model need to be text sequence, it can still be used effectively for the text classification task. We utilized the T5 Model [12] class of Simple Transformer to train a model in multi-class classification setting using T5-base [12] pre-trained model. We used the HuggingFace [15] library for training a classifier using the pre-trained RoBERTa [10] model. We passed the abstracts through the RoBERTa [10] model and took the embedding of the [CLS] token which was passed through a classification head set up to train for multi-class classification.

We decided to perform further experimentation with the RoBERTa [10] model since it had the best performance among the three vanilla model that we built. We created a representation of the abstracts using TF-IDF, and LDA [2] based topic modelling techniques. We combined the LDA [2] and TF-IDF features with the RoBERTa [10] individually and trained two models. We could see the slight improvement in the performance when we combined LDA [2] with the RoBERTa [10] compared to vanilla RoBERTa [10] model. RoBERTa [10] has a limitation of 512 tokens as input. Many of the abstracts have more than 512

---

[8] https://simpletransformers.ai/ (accessed January 25, 2021).

tokens. Thus, the part towards the tail of the abstract was not getting utilized for prediction in these cases. So, we decided to train a sentence level model, where we first tokenized the abstract into sentences using Spacy. These tokenized sentences were assigned the same token as the abstract they were part of. We then trained a classification model using this data using Simple Transformer library with RoBERTa [10] base model. We observed that very short sentences seldom carry enough information to be able to predict the right class just using their constituent words. Hence, we built four models first three by not considering a sentence that had less than 6 and 10 words and the fourth one by taking all the sentences. We found that the model where we had taken sentences with more than 10 words did better than all other models. While scoring these models on the validation data, we scored the individual sentence and averaged the output probability for all the sentences from an abstract. Then, the class with the highest average probability was assigned as the prediction for the abstract. We experimented with the length during the predictions as well and found that taking sentences with more than 5 words tend to perform the best. Finally, we bench-marked our model with one an existing SOTA model which is SciBERT [1].

4.2 Results and Discussion

The results[9] we obtained are presented in Table 2. Here F1 refers to the weighted F1 score. Details relating to models 1, 2, 3 and 4 have been mentioned in the previous section of this paper.

The hyper-parameters corresponding to the best performing versions of RoBERTa [10] models described in Sects. 3.1 (RoBERTa-1), 3.2 (RoBERTa-2), 3.3 (RoBERTa-3), BERT [7] model, T5 [12] model and SciBERT [1] model are mentioned in Table 3. Max. Token is the maximum number of tokens, BS means the batch size and η represents the learning rate.

While performing the experiments, we observed that the Logistic Regression model takes the least amount of training time as compared to other transformer-based models. In each of these models, we saw that the performance was worst for the class Distributed and Cluster Computing (DC) and it was best for the class Computation and Language (CL). On analysing Table 2 we see that the F1 scores are 0.999 on the training set for BERT [7], Model-1, Model-3 and the ensemble model. However, the ensemble model performs best for Validation set (F1 = 0.928). We further observe that zero-shot learning (F1 = 0.114 on the training set and F1 = 0.113 on the validation set) performs worse than all other models. This is because it has not seen the training data. This also confirms that generic models do not perform well on close-ended domain-specific tasks.

---

[9] NOTE: The F1 scores corresponding to the test set have been provided by the SDPRA team after evaluating three of our submissions. Since the number of submissions was restricted to three, we do not have these numbers for the other models which we have developed.

Table 2. Performance of various models. **Bold** highlights the best performing models. Underline denotes existing State Of The Art (SOTA) model.

Model	Train (F1)	Validation (F1)	Test (F1)
Zero shot learning	0.114	0.113	–
BERT	**0.999**	0.898	–
T5	0.977	0.882	–
SciBERT (SOTA)	0.991	0.926	–
Model-1 (RoBERTa)	**0.999**	0.916	–
Model-2 (RoBERTa+LDA)	0.968	0.919	0.912
Model-3 (RoBERTa on sentences)	**0.999**	0.913	–
Model-4 (TF-IDF+Logistic Regression)	0.958	0.915	0.916
Ensemble (Model 1-2-3-4)	**0.999**	**0.928**	**0.929**

Table 3. Hyper-parameters of various models

Model	Max. Token	BS (Train)	BS (Valid)	# Epochs	η	Optimizer
RoBERTa-1	512	32	256	13	0.00002	ADAM [9]
RoBERTa-2	512	32	8	3	0.00002	ADAM [9]
RoBERTa-3	512	8	8	10	0.00004	AdamW [11]
BERT	512	8	8	10	0.00004	AdamW [11]
T5	512	8	8	7	0.00100	ADAM [9]
SciBERT	512	32	256	5	0.00002	ADAM [9]

5   Conclusion and Future Works

Analysing the results mentioned in the previous section, we conclude that the individual models' performances are comparable. We further observe that the ensembling technique outperforms existing SOTA model SciBERT [1] on the validation set in terms of F1.

In future, we would like to experiment by replacing the RoBERTa [10] based embeddings with SciBERT [1] in the sub-systems mentioned in Sects. 3.1, 3.2 and 3.3. Furthermore, we want to study how replacing the majority voting method of Sect. 3.5 with a meta classifier or a fully connected dense layer affects the overall performance. Finally, we shall be working on reducing the model size.

References

1. Beltagy, I., Lo, K., Cohan, A.: SciBERT: a pretrained language model for scientific text. In: Proceedings of the 2019 Conference on Empirical Methods in Natural Language Processing and the 9th International Joint Conference on Natural Language Processing (EMNLP-IJCNLP), pp. 3615–3620. Association for Computational Linguistics, Hong Kong (2019). https://doi.org/10.18653/v1/D19-1371. https://www.aclweb.org/anthology/D19-1371
2. Blei, D.M., Ng, A.Y., Jordan, M.I.: Latent Dirichlet allocation. J. Mach. Learn. Res. **3**(null), 993–1022 (2003)
3. Borrajo, M., Romero, R., Iglesias, E., Marey, C.: Improving imbalanced scientific text classification using sampling strategies and dictionaries. J. Integr. Bioinform. **8**, 90–104 (2011). https://doi.org/10.1515/jib-2011-176
4. Cao, M.D., Gao, X.: Combining contents and citations for scientific document classification. In: Zhang, S., Jarvis, R. (eds.) AI 2005. LNCS (LNAI), vol. 3809, pp. 143–152. Springer, Heidelberg (2005). https://doi.org/10.1007/11589990_17
5. Chang, M.W., Ratinov, L.A., Roth, D., Srikumar, V.: Importance of semantic representation: dataless classification. In: AAAI (2008)
6. Cortes, C., Vapnik, V.: Support-vector networks. Mach. Learn. **20**(3), 273–297 (1995). https://doi.org/10.1007/BF00994018
7. Devlin, J., Chang, M.W., Lee, K., Toutanova, K.: BERT: pre-training of deep bidirectional transformers for language understanding. In: Proceedings of the 2019 Conference of the North American Chapter of the Association for Computational Linguistics: Human Language Technologies, Volume 1 (Long and Short Papers), pp. 4171–4186. Association for Computational Linguistics, Minneapolis (2019). https://doi.org/10.18653/v1/N19-1423. https://www.aclweb.org/anthology/N19-1423
8. Ghanem, M.M., Guo, Y., Lodhi, H., Zhang, Y.: Automatic scientific text classification using local patterns: KDD cup 2002 (task 1). SIGKDD Explor. Newsl. **4**(2), 95–96 (2002). https://doi.org/10.1145/772862.772876
9. Kingma, D.P., Ba, J.: Adam: a method for stochastic optimization (2017)
10. Liu, Y., et al.: Roberta: a robustly optimized BERT pretraining approach (2019). arxiv:1907.11692
11. Loshchilov, I., Hutter, F.: Fixing weight decay regularization in Adam. CoRR abs/1711.05101 (2017). http://arxiv.org/abs/1711.05101
12. Raffel, C., et al.: Exploring the limits of transfer learning with a unified text-to-text transformer. J. Mach. Learn. Res. **21**(140), 1–67 (2020). http://jmlr.org/papers/v21/20-074.html
13. Reddy, S., Saini, N.: SDPRA 2021 shared task data. Mendeley data, v1 (2021). https://doi.org/10.17632/njb74czv49.1. https://data.mendeley.com/datasets/njb74czv49/1
14. Reddy, S.M., Saini., N.: Overview and insights from scope detection of the peer review articles shared tasks 2021 (forthcoming). In: Gupta, M., Ramakrishnan, G. (eds.) PAKDD 2021. LNAI, vol. 12705, pp. 73–78. Springer, Heidelberg (2021)
15. Wolf, T., et al.: Transformers: state-of-the-art natural language processing. In: Proceedings of the 2020 Conference on Empirical Methods in Natural Language Processing: System Demonstrations, pp. 38–45. Association for Computational Linguistics, Online (2020). https://www.aclweb.org/anthology/2020.emnlp-demos.6

The First International Workshop on Data Assessment and Readiness for AI (DARAI 2021)

1st International Workshop on Data Assessment and Readiness for AI

Bortik Bandyopadhyay[4], Sambaran Bandyopadhyay[1(✉)], Srikanta Bedathur[2],
Nitin Gupta[1], Sameep Mehta[1], Shashank Mujumdar[1],
Srinivasan Parthasarathy[3], and Hima Patel[1]

[1] IBM Research AI, Bangalore, India
[2] Indian Institute of Technology Delhi, Delhi, India
[3] The Ohio State University, Columbus, USA
[4] Apple Inc., Cupertino, USA

1 Objectives, Scope, and Contributions

In the last several years, AI/ML technologies have become pervasive in academia and industry, finding its utility in newer and challenging applications. While there has been a focus to build better, smarter and automated AI pipelines, little work has been done to systematically understand the challenges in determining the readiness of data to be fed to this pipeline. Given a business problem, questions whose answers are still elusive include: how does one select the right data from a data source? Is the data collected of the appropriate quality? If not, what cleaning techniques should be applied, and how to determine if the goals of data cleaning are achieved? and so on. Researchers and practitioners alike have increasingly come to the realization that the real-world utility of an ML model is only as good as the data it has been trained on. Therefore, developing techniques and frameworks that help us determine the *readiness of data* for training and deploying machine learning models is of utmost importance.

Readiness of data for AI has gained significant attention in the recent literature. Researchers have used state-of-the-art deep learning approaches on classical problems like missing value imputation [14,15], data cleaning and preparation [3,5], mitigating the effect of outliers [2], fairness considerations including detection of bias [10], explainability of data and machine learning models [1] and the ethical sharing and management of data [7]. Data quality issues can be induced during different phases of data lifecycle starting from its acquisition and integration itself [4]. For example, class imbalance or feature sparsity issues in classification datasets could be a genuine property of the data source or a data acquisition error. We believe that well established metrics are needed to evaluate the readiness of data [6,8], and different metrics may be needed at different stages of data lifecycle to enable automated data preparation modules.

The metrics should be explainable and actionable by the different personas involved in the development of Data and AI lifecycle. For instance, there are data stewards, data integrator, subject matter experts, business users and others who

Point of contact on behalf of the organizing committee.

M. Gupta and G. Ramakrishnan (Eds.): PAKDD 2021 Workshops, LNAI 12705, pp. 117–120, 2021.
https://doi.org/10.1007/978-3-030-75015-2_12

form an integral part of the process and work hand in hand with data scientists. Each of these roles may require different views into the data readiness aspect [9]. For example, we need tools so that a subject matter expert can easily understand the issues detected in the data and give suitable suggestions to add domain knowledge, methods to add this domain knowledge seamlessly into pipeline, and most importantly tools that can make the most of a subject matter expert's time, for example, showing representative samples to the SME, and applying the knowledge to the rest of the dataset [11–13].

The goal of this workshop will be to get researchers working in the fields of data acquisition, data labeling, data quality, data preparation and AutoML areas to understand how the data issues, their detection and remediation will help towards building better models. With the focus on different modalities such as structured data, time series data, text data and graph data, this workshop invites researchers from academia and industry to submit novel propositions for systematically identifying and mitigating data issues for making it AI ready.

1.1 Topics of Interest

Methods of data assessment can change depending on the modality of the data. This workshop will invite submissions for data readiness for different modalities: structured (or tabular) data, unstructured (such as text) data, graph structured (relational, network) data, time series data, etc. We would like to explore state-of-the-art deep learning and AI concepts such as deep reinforcement learning, graph neural networks, self-supervised learning, capsule networks and adversarial learning to address the problems of data assessment and readiness. Following is a (non-exhaustive) list of topics that are of interest to this workshop:

- Algorithms for explainable data quality detection and remediation for ML
- Automated data cleaning workflows with explanations
- Smarter data visualizations for high dimensional data
- Autolabel datasets from small labels of data
- Label noise detection, explanation and incorporating feedback
- Incorporating domain knowledge for data cleaning and data transformations
- Data privacy and encryption techniques, with impact to ML pipeline
- Outlier (or anomaly) detection and mitigation in data
- Detection of bias in data
- Handling corrupted, missing and uncertain data
- Noisy Data Evaluation and Cleaning Recommendation
- Syntactic Data Validations

2   Proposed Workshop Format and Attendees

This workshop can preferably be conducted as a half day meeting (∼3 h 30 min). Following is a preliminary format of the workshop:

- Opening Remarks - 10 min
- Keynote speeches/invited talks (1 talk, 60 min) – 60 min
- Paper presentation (3 papers, 20 min each) - 60 min
- Break - 15 min
- Panel Discussion - 60 min
- Conclusive remarks - 5 min

Our aim is to establish this workshop as a continuing workshop series in top AI and data mining conferences. There were seven research papers submitted to this workshop. Each paper got two reviews. Based on the feedback from the reviewers, we have accepted three papers to be presented in the workshop and to be published in the proceedings. For each accepted paper, both the reviews were positive.

3 Expertise and Experience of the Organizing Committee

The quality assessment and readiness of data is fundamental to data mining. However, formal study of the problem has started very recently. The organizing committee of this workshop has good amount of experience from both academia and industry research. We have published research papers, organized tutorials and demos in AI and data mining conferences (such as KDD 2020, SIGMOD 2020, AAAI 2020, ECAI 2020 and BDA 2020) related to this topic. Members of the organizing committee have experience in different AI applications and with multiple data modalities such as structured (tabular) data, textual data, graph structured data and time series data. Following is the list of the organizing committee members of this workshop.

- Sambaran Bandyopadhyay, IBM Research AI (https://sites.google.com/view/sambaranb/home)
- Hima Patel, IBM Research AI (https://www.linkedin.com/in/patelhima/?originalSubdomain=in)
- Srikanta Bedathur, Indian Institute of Technology, Delhi (https://www.cse.iitd.ac.in/~srikanta/)
- Bortik Bandyopadhyay, Apple, Data Science (http://web.cse.ohio-state.edu/~bandyopadhyay.14)
- Sameep Mehta, IBM Research AI (https://researcher.watson.ibm.com/researcher/view.php?person=in-sameepmehta)
- Srinivasan Parthasarathy, The Ohio State University (http://web.cse.ohio-state.edu/~parthasarathy.2)
- Shashank Mujumdar, IBM Research AI (https://researcher.watson.ibm.com/researcher/view.php?person=in-shamujum)
- Nitin Gupta, IBM Research AI (https://researcher.watson.ibm.com/researcher/view.php?person=in-ngupta47)

References

1. Arya, V., et al.: Ai explainability 360: an extensible toolkit for understanding data and machine learning models. J. Mach. Learn. Res. **21**(130), 1–6 (2020). http://jmlr.org/papers/v21/19-1035.html
2. Bandyopadhyay, S., Lokesh, N., Murty, M.N.: Outlier aware network embedding for attributed networks. In: Proceedings of the AAAI Conference on Artificial Intelligence, vol. 33, pp. 12–19 (2019)
3. Berti-Equille, L.: Learn2clean: optimizing the sequence of tasks for web data preparation. In: The World Wide Web Conference, pp. 2580–2586 (2019)
4. Biljecki, F., Heuvelink, G.B., Ledoux, H., Stoter, J.: The effect of acquisition error and level of detail on the accuracy of spatial analyses. Cartogr. Geogr. Inf. Sci. **45**(2), 156–176 (2018)
5. Heidari, A., McGrath, J., Ilyas, I.F., Rekatsinas, T.: Holodetect: few-shot learning for error detection. In: Proceedings of the 2019 International Conference on Management of Data, pp. 829–846 (2019)
6. Heinrich, B., Hristova, D., Klier, M., Schiller, A., Szubartowicz, M.: Requirements for data quality metrics. J. Data Inf. Qual. (JDIQ) **9**(2), 1–32 (2018)
7. Hirsch, D., et al.: Corporate data ethics: data governance transformations for the age of advanced analytics and AI (initial draft of final report). In: Presented at the Privacy Law Scholars Conference (2020)
8. Hou, Y., et al.: Measuring and improving the use of graph information in graph neural networks. In: International Conference on Learning Representations (2019)
9. Hynes, N., Sculley, D., Terry, M.: The data linter: lightweight, automated sanity checking for ml data sets. In: NIPS MLSys Workshop (2017)
10. Mehrabi, N., Morstatter, F., Saxena, N., Lerman, K., Galstyan, A.: A survey on bias and fairness in machine learning. arXiv preprint arXiv:1908.09635 (2019)
11. Ratner, A., Bach, S.H., Ehrenberg, H., Fries, J., Wu, S., Ré, C.: Snorkel: rapid training data creation with weak supervision. VLDB J. **29**(2), 709–730 (2020)
12. Raza, M., Gulwani, S.: Automated data extraction using predictive program synthesis. In: Thirty-First AAAI Conference on Artificial Intelligence (2017)
13. Rekatsinas, T., Chu, X., Ilyas, I.F., Ré, C.: HoloClean: holistic data repairs with probabilistic inference. Proc. VLDB Endow. **10**(11), 1190–1201 (2017)
14. Wu, R., Zhang, A., Ilyas, I.F., Rekatsinas, T.: Attention-based learning for missing data imputation in HoloClean. In: Proceedings of Machine Learning and Systems, pp. 307–325 (2020)
15. Yoon, J., Jordon, J., Schaar, M.: Gain: missing data imputation using generative adversarial nets. In: International Conference on Machine Learning, pp. 5689–5698 (2018)

Cooperative Monitoring of Malicious Activity in Stock Exchanges

Bhavya Kalra$^{(\boxtimes)}$ ⓘ, Sai Krishna Munnangi ⓘ, Kushal Majmundar ⓘ,
Naresh Manwani ⓘ, and Praveen Paruchuri ⓘ

International Institute of Information Technology, Hyderabad, Hyderabad, India
{bhavya.kalra,krishna.munnangi}@research.iiit.ac.in,
kushal.majmundar@students.iiit.ac.in,
{naresh.manwani,praveen.p}@iiit.ac.in

Abstract. Stock exchanges are marketplaces to buy and sell securities such as stocks, bonds and commodities. Due to their prominence, stock exchanges are prone to a variety of attacks which can be classified as external and internal attacks. Internal attacks aim to make profits by manipulation of trading processes e.g., Spoofing, Quote stuffing, Layering and others, which are the specific focus of this paper. Different types of proprietary fraudulent activity detectors are deployed by stock exchanges to analyze the time series data of trader's activities or the activity of a particular stock to flag potentially malicious transactions while human analysts probe the flagged transactions further. The key issue faced here is that while the number of anomalous transactions identified can run into thousands or tens of thousands, the number of such transactions that can realistically be probed by human analysts would be a small fraction due to resource constraints. The issue therefore reduces to a dynamic resource allocation problem wherein alerts that represent the most malicious transactions need to be mapped to human analysts for further probing across different time intervals. To address this challenge, we encode the scenario as a Cooperative Target Observation (CTO) problem wherein the analysts (modeled as observers) perform a cooperative observation of alerts that represent potentially malicious activity (modeled as targets) and develop multiple solution approaches in order to identify malicious activity.

1 Problem Statement

A stock exchange is a facility where stock brokers and traders can buy and sell securities such as stocks, bonds, commodities, futures and other financial instruments. Buyers and sellers submit their trade orders during the exchange operating hours abiding to the rules and regulations imposed by the exchange. Stock exchanges maintain **order book** in which the buy/sell orders are anonymized and recorded. There are different type of orders that a stock exchange provides,

---

B. Kalra and S. K. Munnangi—Equal contribution.

© Springer Nature Switzerland AG 2021
M. Gupta and G. Ramakrishnan (Eds.): PAKDD 2021 Workshops, LNAI 12705, pp. 121–132, 2021.
https://doi.org/10.1007/978-3-030-75015-2_13

that are either instantaneously executed or retained until specific conditions are met. The order matching system periodically tries to match buy and sell orders and if there is a match, the trade is executed i.e., the transaction between the traders takes place [8]. Two orders are said to be matched if one of them is a buy order while the other is a sell order and: (a) The buying price is greater than or equal to selling price (timestamp of the order is used as a tie breaker) and (b) If the buying and selling volumes match, the transaction is completed between the buyer and seller and both the orders are marked as completed and removed from order book. (c) If the volumes do not match, one of the orders is marked executed while the other is deemed partially executed and remains in the order book with the new volume set to the difference between the buying and selling volumes.

Due to their prominence as barometers of economy and the enormous amount of wealth that gets created, lost or traded, stock exchanges face a great amount of scrutiny and are prone to a variety of external and internal security attacks [10,17]. While Cyber security attacks, DDoS attacks, Physical security attacks and others are common types of external attacks [2,7,14] that aim to bring the exchange down, internal attacks aim to make profit by manipulating the market and trades. For purposes of this paper, we focus specifically on **internal attacks** [4,9,13] such as Pump and dump, Layering, Wash Trading and others. To tackle the different types of internal attacks, we encode the scenario as a Cooperative Target Observation (CTO) problem and develop multiple solution approaches wherein the analysts modeled as observers perform a cooperative observation of this potential malicious activity (i.e., malicious trades and traders).

There are a number of players in a stock exchange including individual traders, stock brokers, high frequency traders (HFTs) and many other financial institutions that result in millions of transactions per day. All these transactions result in two forms of structured time series data: (i) data based on each trader's activity across a particular stock or multiple stocks and (ii) data of all the traders activity across a particular stock. Both the forms of the data record high volumes of transactions and the primary focus of this paper is on the second form of data where analyzing each trader's activity coupled with other traders activity across a particular stock is easier. The recorded historical data can provide valuable information as explored for the CTO problem in [21]. However, due to the high volume of transactions, it is significantly hard for an exchange to analyze all the order requests placed and track individual trader activity to check and identify malicious transactions. Different types of proprietary fraudulent activity detectors [6,19] are deployed by stock exchanges to flag malicious transactions. However, these detectors typically flag the potential malicious transactions as alerts while human analysts are needed to probe these transactions further to validate their genuineness and identify the appropriate action to take [5]. Though, the number of alerts generated can quickly become intractable due to resource constraints.

The problem can therefore be decoupled into the following two sub problems to do the following:

- **Alert Generation:** Detection of potentially malicious trades along with computation of a severity score to represent the maliciousness.
- A **dynamic resource allocation** problem wherein alerts that represent the most malicious trades need to be mapped to human analysts for further probing across the different time intervals.

2   Cooperative Target Observation (CTO)

Formally as defined in [15], the CTO problem is a tuple $<S,\ O,\ X>$ where **S** is a two-dimensional, bounded, enclosed spatial region, **O** is a team of k_1 observer agents/robots with observation sensors of limited range (denoted by *sensor_range*) and **X** is a set of k_2 targets. The goal is to maximize

$$\sum_{t=0}^{T}\sum_{j=1}^{k_2}\bigvee_{i=1}^{k_1} a_{ij}(t),\ \text{given}\ \bigcup_{o_i\in O} sensor\_range(o_i) \ll S$$

where $a_{ij}(t)$ is 1 if agent/robot o_i is monitoring target x_j at time t and is 0 otherwise and $\bigvee$ is the logical OR operator.

[11] presents a K-means based solution for a decentralized version of the CTO problem, where each observer acts independently to maximize its own observation of targets. The CTO problem is defined for a physical space with physical location coordinates as key parameters while players in a stock exchange operate in a **behavioral space**. We therefore map the CTO tuple $<S,\ O,\ X>$ to a behavioral space as follows:

1. **S** - Models the transaction space i.e., captures each and every transaction made by all the traders operating within the exchange.
2. **O** - Human analysts are the observers, where the bandwidth of the analyst corresponds to the sensor range of the observer.
3. **X** - Potentially malicious transactions flagged as alerts are the targets.

In behavioral space the trades by default don't get assigned to an analyst and hence the first step would therefore be to perform an assignment wherein we assign the set of flagged trades to different human analysts based on their skill set. The next step is to prioritize the assigned trades for each analyst based on the severity score and level of match.

2.1   Generation of Alerts

A generated alert is a tuple consisting of the potentially malicious transaction along with a severity score identified through anomaly detection. The idea of anomaly detection is to detect abnormal data points i.e. trades which are unusual or different from normal trades. While there can be multiple ways to model an anomaly detector [3], we use the one class SVM (OCSVM) [12] which is a popular technique used extensively in the anomaly detection literature. In particular, we train a SVM to learn a classification function that is negative for regions with small densities of data points and positive for regions with high density of data points. We can obtain the optimal classification function $f(x)$ by experimenting with two different types of kernels namely Linear and RBF kernel.

Feature Extraction: A key step to perform anomaly detection is to do feature extraction. A transaction is a buy/sell request made by the trader at a particular time instance. In our case the transactions occur at micro-second level. We define a **trade** as a collection of transactions taking place over a particular unit of time, say a second. We begin the process of feature extraction by acquiring pairs of <*trader, timestamp*> for a trade. These pairs uniquely identify a trade made by a trader and hence act as an index for our trades. We extract a total of 32 features for each pair of <*trader, timestamp*> and the further details are omitted due to space constraints.

Severity Score: The classification function provides a classification score and we classify a trade as non malicious if this value is above 0 and malicious if value is below 0 (Note that the threshold value 0 can be calibrated according to the data). The classification score acquired from classification function of OCSVM is unbounded so we transform the score using function z, where:

$$z(x, \theta) = \frac{1}{1 + e^{-(\theta - f(x))}}$$

θ in the transformation function is the threshold for the classification function $f(x)$. The transformed value is in the range 0 to 1 and is referred to as severity score. The larger the value, the more malicious the trade is. Therefore, the most malicious trades are closer to 1, while non-malicious trades are closer to 0. This severity score is then appended to the existing feature tuple for each trade. Thus, each tuple for a trade is in the form <***trade_id, trader_id, feature_vector, severity score, ground truth***>. Collection of all the tuples is denoted by **B**.

2.2   Trade Level Alerts Generation

The output of alerts generation module is a batch of alerts B which is divided into several mini batches $b_t \forall t \in \{1 \dots r\}$ such that each mini batch contains T tuples. The mini batches are ordered by their starting time and are processed in iterations. We begin by clustering the trades of a mini batch into M clusters (to ensure equal distribution of alerts among the analysts) based upon only the feature vectors present in the tuple, where M denotes the number of analysts available. To analyze the various batches of trades, we use an iterative K-means approach which has the following steps:

1. For mini batch b_t with trades starting at time t, clustering using K-means is performed to obtain M centroids $(C_{b_t}^1 \dots C_{b_t}^M)$. The clustering is started with $(C_{b_{t-1}}^1 \dots C_{b_{t-1}}^M)$ as cluster centers for all mini batches following first mini batch (i.e., cluster center from previous time step).

2. Tuples in each cluster are classified as most malicious based on a threshold on severity scores ($threshold_s$) and then the top tuples are chosen for further processing by human analysts selected using algorithms discussed in the next subsection.

3. Steps 1 and 2 are performed iteratively for each mini batch.

Note: For the first iteration i.e. for the first mini batch, the cluster centers are randomly generated.

The K-means clustering and the process of prioritizing tuples (i.e., choosing the most malicious tuples) are the two important steps in the process of filtering out the non-malicious tuples. We implemented the weighted K-means based on Mahalanobis distance between feature vectors and weighed by the severity scores of each trade. Here, the cluster is allotted according to $dist(x, C_j) \times weight[x]$ where x denotes the feature vector and C_j denotes the jth cluster center. The K-means clustering was followed by three approaches to pick the top malicious tuples. The three approaches we present below to detect the top malicious tuples, are provided the clusters based on feature vectors as well as the severity score, trader and ground truth associated with each trade. Following are the details of the approaches:

1. Top-N Method: For each cluster, we first remove the tuples that have severity value less than $threshold_s$ to ensure that we pick only those which might be malicious. We then pick a maximum of top N tuples, from the list of tuples sorted in descending order of severity value. This process is repeated for each cluster during each iteration (corresponding to each mini batch b_t) as part of Step 2 of the Iterative K-means approach. Next, the top tuples from the clusters are assigned to one of the M analysts for further analysis.

2. Eligibility Traces based Method: In the trading domain, past severity values and occurrence of malicious trades by specific traders or in specific circumstances in the past, also help to identify with better accuracy the maliciousness of a trade. Hence, we introduce the notion of eligibility traces to capture past information during decision making which has its origin in reinforcement learning literature [20]. From the context of CTO literature, eligibility traces based method is a new solution technique being introduced in this paper.

In this method, we initialize the values of eligibility scores and the frequency of each trader to be zero for the first iteration. In subsequent iterations, the eligibility scores and the frequency of appearance of each trader are updated and reused in next iteration. The entire process of choosing N tuples from the cluster A using eligibility traces is presented in Algorithm 1. We can also alter v_1 and v_2 in the algorithm and make it favour exploitation or exploration.

3. Explore-Exploit Method: Inline with CTO solution presented in [1], we build a explore-exploit based model to ensure that when there are iterations where we are unable to capture malicious behaviour, we explore more with the expectation to catch malicious behaviour (not caught by the anomaly detector). To model this approach, we pick 70% of values based on the top-N approach as presented above while for the remaining 30%, we randomly pick them from the transactions not picked by top-N and have severity values above $threshold_s$.

Algorithm 1. Eligibility Traces based Method

Require: transaction-tuples (A) in a cluster, N - the number of tuples to be chosen
1: transaction tuples to be analyzed, $H \leftarrow \phi$
2: **for** i in $1 \dots N$ **do**
3:     $max\_val \leftarrow 0$; $top\_tuple \leftarrow 0$
4:     **for each** tuple $<t_j, s_j>$ in A **do**
5:         $value \leftarrow (eligibility[t_j] + freq[t_j] + 1) \times s_j$
6:         **if** $value \geq max\_val$ **then**
7:             $max\_val \leftarrow new\_val$; $top\_tuple \leftarrow <t_j, s_j>$
8:         **end if**
9:     **end for**
10:     Remove top_tuple from A and add to H
11:     **for** each trader t_j **do**
12:         **if** t_j is the trader in top_tuple **then**
13:             $eligibility[t_j] \leftarrow v_1 \times (eligibility[t_j] + freq[t_j] + 1) \times s_j$
14:         **else**
15:             $eligibility[t_j] \leftarrow v_2 \times eligibility[t_j]$
16:         **end if**
17:     **end for**
18: **end for**
19: Update frequencies of each trader by number of times of their appearance in A
return H

2.3 Trader Level Alerts Generation

We map the malicious trader detection to a CTO problem, where the analyst is the observer trying to analyze the traders (i.e. targets) in the domain. Output of alerts generation module is a trade-tuple for each trade which we divide into batches of tuples corresponding to T trades as earlier.

Eligibility Score: Since eligibility scores are a measure of eligibility of a trader to be subjected to further analysis, we use these scores to rank the traders in the order of their maliciousness instead of performing a trade level analysis. The eligibility scores obtained over several mini-batches ($b_t s$) are used for this purpose, where the traders with higher scores are ranked higher i.e., trader with highest eligibility score is ranked One.

Average Rank: In each iteration, we order the traders based on their eligibility scores and allot them ranks. These ranks are recorded for every iteration. At the end of last iteration, we take an average of the recorded ranks for each trader. This results in an order of priority for human analysts.

3 Experiments

3.1 Data Generation

We used a proprietary data generation software owned by a private firm to generate synthetic trading data. The data generator was capable of generating trading

Table 1. Datasets

Datasets	Transactions	Total trades	Non malicious trades	Malicious trades	Time (hrs)	Number of attacks	Type of attacks
Dataset 1	100000	37646	28033	9613	7	1	L
Dataset 2	100000	18575	16667	1908	3	1	PnD
Dataset 3	100000	84676	78464	6212	19	1	WT
Dataset 4	100000	53261	48222	5039	9	2	PnD + WT
Dataset 5	100000	39142	33672	5470	6	3	PnD + WT + L

Table 2. Performance of alert generator

Datasets	ν		γ		Threshold		$Recall_{ag}$		$Precision_{ag}$		Accuracy	
	Linear	rbf	Linear	rbf	Linear	rbf	Linear	rbf	Linear	rbf	Linear	rbf
Dataset 1	0.03	0.03	1	0.1	0.00019	6.65	97	95	77	65	80	65
Dataset 2	0.03	0.03	1	0.1	0.0001	5.56	94	96	36	36	36	35
Dataset 3	0.03	0.03	1	0.1	0.00019	39.0	90	95	27	34	28	47
Dataset 4	0.03	0.03	1	0.1	0.00027	14.65	94	98	36	54	39	55
Dataset 5	0.03	0.03	1	0.1	0.00024	8.6	95	99	53	54	52	54

data for stocks from different institutions and could be configured to inject a variety of - three different attack types (or combinations of those). Along with the **attack types** other configurable parameters include **number of malicious and non-malicious traders, the time interval for which malicious trading data is to be generated** and **the total volume of transactions between the traders.** The generator outputs two datasets - one dataset contains all the transactions while the other contains the list of malicious transactions.

3.2 Attack Types

We generate data featuring the three different attacks types, namely Layering (L), Wash Trading (WT) and Pump and dump (PnD).

1. **Layering:** Layering [18] is the act of placing multiple visible non-bonafide orders with the intent of creating a false impression of supply or demand.
2. **Wash Trading:** A wash trade [18] is a form of market manipulation in which an investor simultaneously sells and buys the same financial instruments to create misleading, artificial activity in the marketplace.
3. **Pump and dump:** Pump and dump [18] is a scheme that attempts to boost the price of a stock through recommendations based on false, misleading or greatly exaggerated statements.

Datasets: We performed our experiments on 5 different datasets: The first three datasets are single attack datasets with the above mentioned attacks. The fourth dataset includes Pump and dump and Wash Trade attacks, while the fifth dataset comprises of all three i.e., Pump and dump, Wash Trade and Layering.

Each dataset contains approximately 100K buy/sell transactions made over a period of several hours. Details of each dataset are presented in Table 1. The number of malicious transactions are much smaller than the number of non malicious ones. One unit of time is set to 1 s and the transactions occur at a micro second level which map to several thousands of trades.

3.3   Parameters for Alerts Generation Module

The large batch B is fed as input to the alerts generation module. Anomaly detection process is initiated to compute severity score and proceeds as follows:

1. Training: The process of distinguishing malicious and non-malicious trades begins with training the OCSVM. The batch B is split into non-malicious and malicious trades (using ground truth provided by the data generator which has the list of transactions marked malicious). We train our OCSVM on 80% of the non malicious trades. The following parameters affect how well our trained OCSVM will perform: (a) the type of kernel - linear or gaussian (rbf) (b) the regularization parameter ν (c) the parameter γ

We train using both the linear and rbf kernels and perform Grid search [16] to identify optimal values for parameters ν and γ for both the scenarios. The values for the parameters are considered optimal when the value for $Recall_{ag}$ is high, where $Recall_{ag}$ for alert generator is defined as the ratio of malicious trades identified out of all malicious trades in the given batch B. The parameters obtained for such setting catch most of the malicious trades even if we err on side of being cautious i.e. non-malicious trade getting classified as malicious. The search space for γ is $[2^{-4}, 2^4]$ while search space for ν is selected as $[0.01, 0.8]$ with step size 0.01. The resulting metrics such as $Recall_{ag}$, $precision_{ag}$ and Accuracy for alert generator have been stored in Table 2.

2. Testing: We prepare our test data by taking 20% of the non malicious data i.e. data not used for training and combine it with entire of the malicious data. We measure performance of the detector on the test data by plotting $Recall_{ag}$ vs false positive rate i.e. the Receiver Operator Characteristics (ROC) curve as shown in Fig. 1a. We define the false positive rate as the ratio of non malicious trades misclassified as malicious amongst all trades classified as malicious. We determine threshold for our classifier based on corresponding highest area under the curve and recall combination values. We observe that rbf kernel performs better than linear kernel for most datasets (Table 2).

3.4   Malicious Trader Identification

The following parameters were provided to the algorithm: (a) Number of trades to be picked (default 5) (b) Number of clusters (default 5) and (c) Severity Threshold (default 0.5).

We measure the outcome of trade selection on the basis of recall and precision values. The eligibility traces algorithm has performed better than all the other

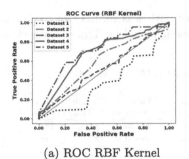

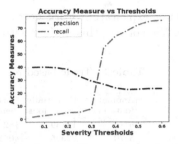

(a) ROC RBF Kernel   (b) $Recall_{dra}$ vs $Precision_{dra}$ Trade Off

Fig. 1.

approaches in terms of $Recall_{dra}$. However, it performs poorly on $Precision_{dra}$ which is acceptable as we are able to catch most malicious trades. There is a trade off between $Precision_{dra}$ and $Recall_{dra}$, which can be demonstrated by plotting $Recall_{dra}$ and $Precision_{dra}$ vs severity threshold, evident in Fig. 1b.

We use three datasets for malicious Trader identification results. The Trader dataset 1 has four out of seven traders as malicious, while dataset 2 has two out of seven traders as malicious. The third dataset however has 5 out of 100 traders as malicious. Each trader dataset has trades with multiple (3) type of attacks similar to dataset 5 (Table 1). As shown in Table 4, the eligibility traces algorithm provided the best results over all datasets. We build upon trades identified by the algorithm to identify the most malicious traders based on their eligibility. For each mini batch b_t we rank the malicious traders using eligibility and take their average rank across all the mini batches. The parameters (values) set as v_1 = 0.25 and v_2 = 0.9 support exploration in the eligibility traces algorithm. The same algorithm can be altered to support exploitation by setting values $v_1 > 1.0$ and v_2 closer to 0. In Table 4, we compare results of these two approaches on different datasets. We introduce another accuracy metric(η) which follows the formula

$$\eta = \left(1 - \sum_{i=1}^{m} \frac{|pr_i - tr_i|}{total\ number\ of\ traders}\right) \times 100$$

where pr, tr represent average predicted rank and true rank respectively and m is the number of malicious traders.

Summary of Results: The results for trade based algorithms on multiple attack dataset are included in Table 3. We find that the eligibility traces algorithm outperforms the other two variants. We also observe that by varying the M and N, the precision and recall values follow an inverse relationship. This relationship is shown in Fig. 1b, where each point represents precision and recall for the pair of number of analysts and number of traders. As the number of traders and number of trades picked in each b_t are increased, recall goes low and precision goes high. It is observed that parameters supporting exploration work better when more malicious traders are involved in trading while parameters supporting exploitation work better in datasets with less number of malicious traders. We confirm this finding with the results achieved on Trader dataset 3.

Table 3. Trade based results on Dataset 5 (N = 8, M = 8)

Algorithm	Anomalies (dataset)	Potentially anomalous	Correctly identified	False negatives	False positives	Recall (R_{dra})	Precision (P_{dra})
Top-n	5470	15739	1824	3646	13915	33	12
Eligibility	5470	22955	4055	1415	18900	**74**	18
Explore-Exploit	5470	16164	1927	3543	14237	35	12

Table 4. Trader based results on Dataset 5 (N = 5, M = 5)

Dataset	Malicious traders	Explore score (η)	Exploit score (η)
Dataset 1	4	85.71	71.44
Dataset 2	4	71.44	100
Dataset 3	4	36	94

4 Conclusions and Future Work

Aim of this paper is to improve monitoring of malicious activity in a stock exchange. To address this challenge, we encoded the scenario as a Cooperative Target Observation (CTO) problem and then introduced three different solution approaches to perform better observation of malicious trades and subsequently malicious traders. From the context of CTO literature, Eligibility Traces based method is a new solution approach being introduced in this paper. We also developed a simple alerts generation module that uses OCSVM for testing the performance of our solution. Our results show that: (a) There is a trade-off between recall and precision and we try to maximize recall in order to optimize the capture of malicious traders and trades. (b) Eligibility Traces based method works best achieving both higher recall and precision compared to the other methods for CTO. (c) While single attack types are easier to handle, it becomes tougher to identify malicious traders/trades as more attack types get involved. However, we are able to identify accurately, nearly all the malicious traders across all datasets. As part of future work, given the open nature of a marketplace such as stock exchange, we plan to investigate Reinforcement Learning based approaches to perform dynamic resource allocation.

Acknowledgements. We would like to thank CognitiveScale for providing access to their data generation software for experimentation purposes and related discussions and for the generous support provided.

References

1. Aswani, R., Munnangi, S.K., Paruchuri, P.: Improving surveillance using cooperative target observation. In: Thirty-First AAAI Conference (2017)
2. Brandom, R.: Hackers increasingly turn their sights on stock exchanges (2013). https://www.theverge.com/2013/8/1/4578688/hackers-turn-their-sights-on-stock-exchanges
3. Chandola, V., Banerjee, A., Kumar, V.: Anomaly detection: a survey. ACM Comput. Surv. (CSUR) **41**(3), 1–58 (2009)
4. Frunza, M.C.: Introduction to the Theories and Varieties of Modern Crime in Financial Markets. Academic Press, Cambridge (2015)
5. Ganesan, R., Jajodia, S., Shah, A., Cam, H.: Dynamic scheduling of cybersecurity analysts for minimizing risk using reinforcement learning. ACM Trans. Intell. Syst. Technol. (TIST) **8**(1), 4 (2016)
6. Goldberg, H.G., Kirkland, J.D., Lee, D., Shyr, P., Thakker, D.: The NASD securities observation, new analysis and regulation system (SONAR). In: IAAI, pp. 11–18 (2003)
7. Goodin, D.: How elite hackers (almost) stole the NASDAQ (2014). https://arstechnica.com/information-technology/2014/07/how-elite-hackers-almost-stole-the-nasdaq
8. Harris, L.: Trading and Exchanges: Market Microstructure for Practitioners. OUP, Oxford (2003)
9. Imisiker, S., Tas, B.K.O.: Wash trades as a stock market manipulation tool. J. Behav. Exp. Finance **20**, 92–98 (2018)
10. Ingraham, N.: The US government has charged five men with the largest hacking scheme in us history (2013). https://www.theverge.com/2013/7/25/4556594/five-men-charged-in-largest-hacking-scheme-in-us-history
11. Luke, S., Sullivan, K., Panait, L., Balan, G.: Tunably decentralized algorithms for cooperative target observation. In: Proceedings of the Fourth International Conference on Autonomous Agents and Multiagent Systems, pp. 911–917. ACM (2005)
12. Manevitz, L.M., Yousef, M.: One-class SVMs for document classification. J. Mach. Learn. Res. **2**, 139–154 (2002)
13. Montgomery, J.D.: Spoofing, market manipulation, and the limit-order book (2016). https://ankura.com/insights/spoofing-market-manipulation-and-the-limit-order-book
14. Nikkei: Malaysia probes cyberattack suspected of disrupting online stock trading (2017). https://asia.nikkei.com/Markets/Equities/Malaysia-probes-cyberattack-suspected-of-disrupting-online-stock-trading
15. Parker, L.E.: Distributed algorithms for multi-robot observation of multiple moving targets. Auton. Robots **12**(3), 231–255 (2002)
16. Pedregosa, F., et al.: Scikit-learn: machine learning in Python. J. Mach. Learn. Res. **12**, 2825–2830 (2011)
17. Rashid, F.Y.: Cyber attacks against stock exchanges threaten financial markets (2013). https://www.securityweek.com/cyber-attacks-against-stock-exchanges-threaten-financial-markets-report
18. Saeedi, A., Hamedi, M.: Financial Literacy: Empowerment in the Stock Market. Springer, Cham (2018). https://doi.org/10.1007/978-3-319-77857-0
19. SMARTS: Nasdaq (smarts) trade surveillance for automated capture and monitoring of all trading activity. https://new.nasdaq.com/solutions/nasdaq-trade-surveillance

20. Sutton, R.S., Barto, A.G.: Reinforcement Learning: An Introduction. MIT Press, Cambridge (2018)
21. Wang, J., Cao, J., Wang, S., Yao, Z., Li, W.: IRDA: incremental reinforcement learning for dynamic resource allocation. IEEE Trans. Big Data 1 (2020). https://doi.org/10.1109/TBDATA.2020.2988273

Data-Debugging Through Interactive Visual Explanations

Shazia Afzal[1(⊠)], Arunima Chaudhary[1,3], Nitin Gupta[1], Hima Patel[1],
Carolina Spina[2], and Dakuo Wang[3]

[1] IBM Research, New Delhi, India
{shaafzal,ngupta47,himapatel}@in.ibm.com
[2] IBM Argentina, Buenos Aires, Argentina
carolina.spina@ibm.com
[3] IBM Research, Cambridge, USA
{arunima.chaudhary1,dakuo.wang}@ibm.com

Abstract. Data readiness analysis consists of methods that profile data and flag quality issues to determine the AI readiness of a given dataset. Such methods are being increasingly used to understand, inspect and correct anomalies in data such that their impact on downstream machine learning is limited. This often requires a human in the loop for validation and application of remedial actions. In this paper we describe a tool to assist data workers in this task by providing rich explanations to results obtained through data readiness analysis. The aim is to allow interactive visual inspection and debugging of data issues to enhance interpretability as well as facilitate informed remediation actions by humans in the loop.

Keywords: Data readiness · Data quality · Visual analytics · Explainability · Interactive data debugging · Human-in-the-loop

1 Introduction

Good quality data is a prerequisite for building efficient machine learning models. Anomalies or inconsistencies in training data can severely impact model complexity as well as accuracy. This issue is more crucial as data is being created and ingested from various sources, in different formats at a remarkable velocity. As a result, data remains susceptible to errors or irregularities that may be introduced during collection, aggregation or annotation stages. This necessitates profiling and assessment of data to understand its suitability for machine learning tasks. We refer to the set of various data profiling, quality assessment operations and the associated remediations as *Data Readiness Analysis* to imply the readiness of a given dataset before it enters an AI pipeline. Figure 1 shows the positioning of data readiness assessment in a typical machine learning workflow and the

---

The first two authors have contributed equally to this paper.

© Springer Nature Switzerland AG 2021
M. Gupta and G. Ramakrishnan (Eds.): PAKDD 2021 Workshops, LNAI 12705, pp. 133–142, 2021.
https://doi.org/10.1007/978-3-030-75015-2_14

interaction of various personas during this process. The primary objective of conducting readiness analysis on raw data is to get an estimate of data quality issues upfront so that data workers like data stewards, data scientists, subject matter experts, or machine learning scientists can get relevant data insights and take remedial actions to rectify any issues. The figure illustrates how raw data after ingestion is analysed for quality checks involving human review, as required, before entering a machine learning pipeline.

In addition to the typical data validation checks like correlation detection, missing values detection, presence of duplicates, flagging of outliers, and checking for format inconsistencies, data readiness analysis includes more advanced data profiling operations relevant to machine learning contexts like checking for class imbalance, identification of mislabelled samples, collinearity checks, presence of discriminatory features, occurrence of bias and detection of overlapping points. Such an assessment is accompanied with appropriate remediations or transformations to be performed on data to rectify the issues identified. This often involves a human in the loop in the form of a subject matter expert, a data steward, or a data scientist who has to understand, validate or apply suitable modifications. Human supervision is desirable for carrying out such data modifications over entirely automatic remediations of any required edits.

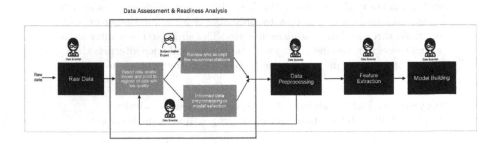

Fig. 1. Data readiness analysis in a typical machine learning pipeline.

This is because data is a critical asset to any organization, and any changes or transformations that may impact its characteristics or utility need to be well-tracked and inspected. The human effort required for validating the data remedial operations is easily the most expensive, tedious and time-consuming factor in the data readiness workflow. Efforts to make this task easier and faster can drastically reduce the cost and time expected for data preparation. Data owners and practitioners ought to be provided with meaningful insights into the output of the data readiness evaluation. This is even more applicable for high dimensional and predominantly numerical datasets where only table views may not be so intuitive. It should be the responsibility of data readiness analysis to help users understand and inspect any data issues prevalent or detected by the underlying analysis techniques and describe the reasoning behind the decisions or insights provided. One way to achieve this is to include meaningful and appropriate explanations of results for end-user review. Explaining the logic and functioning of underlying methods to the users brings transparency and builds trust. Similarly, when users are able to understand and explore the results of data readiness analysis, they are empowered to take more informed decisions regarding application of remedial or suggested actions. This improves their trust and perception of the analysis performed and eventually ensures that data remediation is conducted in a more reliable and less error-prone manner. Offering such kind of interpretability is consistent with the agenda of Explainable Artificial Intelligence (XAI) that seeks to generate explanations and describe the reasoning behind machine-learning decisions and predictions to support user understanding and comprehension [8]. There is compelling evidence to indicate the effectiveness of user interactivity with AI-systems [1] as well as the use of visual analytics in driving and aiding such interactions [5].

Inspired by this critical line of work and realizing a similar gap in the data preparation phase of the AI pipeline, we are building an explainability framework to enable *Data-Debugging through Interactive Visual Explanations* as a supplement to our efforts in data assessment and readiness analysis [6]. As data readiness is an umbrella term we use to refer to all operations related to data profiling, cleaning and quality assessment we review relevant literature in the following Sect. 2 for setting the right context. However, our focus is mainly on quality operations and remediations relevant to machine learning tasks which is why our framework is distinct in terms of the data assessment dimensions included and user operations permitted. We describe this framework in Sect. 3.

2   Related Work

Various commercial tools exist that give powerful visualizations of common data issues like outliers, missing values, duplicates, etc. Tableau and Microsoft Power BI are some of the most widely adopted visualization tools in the industry. They are easy to use with their drag and drop functionalities and cloud support but tend to have overlapping features. They primarily depend on users to pick or generate optimal parameters for visual discovery and interpretations.

Tools such as Wrangler and OpenRefine offer customizable data transformations further aiding in the data mining process. Wrangler offers an interactive system for creating data transformations. It combines direct manipulation of visualized data with automatic inference of relevant transforms, enabling analysts to iteratively explore the space of applicable operations and preview their effects [7]. OpenRefine prides itself for turning messy data into usable, clean data. Users can explore data to see the big picture, clean and transform data, and reconcile data with various web services [4]. While OpenRefine cleans data, Wrangler provides means to reorganize data in order to create easy to understand views. However, both the systems lack at providing any data quality parameters or their explanations and once again depend on the user to drive their own ML processes to generate and interpret the views.

Google facets is yet another effort towards putting across a holistic picture of data at different granularities but lacks at providing data manipulation tools to gain further insights. Tensorflow Embedding Projector moves a step forward as it offers an advanced tool for interactive visualization and analysis of high-dimensional data using ML techniques such as UMAP, TSNE and PCA. It provides multiple dimensions of interpretation: Exploring local neighborhoods, Viewing global geometry and finding clusters and Finding meaningful "directions" but once again does not go beyond providing structural insights of the data distribution [11].

We don't find much related work where systems generate views geared towards explaining data quality issues specific to ML use-cases like label noise, class overlap, etc. and providing useful interpretations for the same. Tools like explAIner [12] and TELEGAM [5] that combine interactive and visual analytics capabilities for explainable and interpretable analysis of machine learning models, serve as a motivating example for our work which is a novel effort in the data readiness analysis space. We anticipate that such a tool will improve the interpretability of quality metrics and make the functioning transparent.

3   Data-Debugging Through Interactive Visual Explanations

In this section we introduce an actionable framework for data debugging that allows inspection of results obtained through data readiness assessment using interactive visual explanations. Even though we consider structured or tabular data as a use case here, the principles are scalable to other data formats like unstructured or time-series data-types. The main objective here is to provide human understandable explanations in an integrated manner. We combine elements of visual, textual and tabular data views to create comprehensive explanations of the readiness outputs. The explanations are provided at two levels - a global level explanation to summarise the results of each dimension or metric, and an instance level local explanation to allow inspection of individual data points. The two levels are provided to cater to the requirements and expertise of the target end user and their specific needs goals.

Both visual and textual aids are used to design supporting explanations so that it is easier for users to explore and understand the data issues flagged during data readiness analysis. Using visualization techniques is a quick and intuitive way of communicating results and illustrating complex information. Human beings can process visual information faster and more easily which is why data visualization has been traditionally been a powerful means of uncovering insights from data. Charts and maps of varying complexities are commonly available in commercial data analysis tools. These tools generally have some form of a dashboard view customized to relevant data analytics for the intended audience. Incorporating visualizations in the readiness explanations is therefore an obvious choice. However, without a supplementary narration, solely graphical or visual displays are open to subjective interpretation. It is possible for end-users, especially novices, to derive unintended or incorrect data insights from unaided visual displays. To facilitate accurate understanding, we therefore provide accompanying textual explanations in natural language. The purpose is to explicitly call out the intended insight objectively and leave no room for misunderstanding or deception. Moreover, combination of visualization and verbalization techniques has been found to be a powerful strategy in explainable ML [10] which is why we are confident of applying the same concept to the data readiness domain.

We also provide a mapping of the visualization to the actual records in a tabular format. This has two purposes. One, is to facilitate an integrated view of the data with the visualization so that the end-user retains the context of the data space. Two, is to assist in decision-making with respect to relevant remediations by permitting a scrutiny of actual feature distributions reflected through the tabular data view. This is important when data is plotted into a lower-dimensional space for better viewing but the corresponding remedial action needs a careful consideration of the actual feature values. Figure 2 illustrates the expected sequence of user interactions when analysing results through this tool.

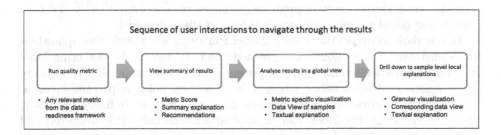

Fig. 2. Sequence of steps a user may go through during interaction with the tool

Overall, we believe that providing a comprehensive explanation of data readiness analysis should include multiple modalities so that users get a complete perspective on the data issues detected. We anticipate the following, not exhaustive, use-cases for such a tool support:

– Data scientists can use the tool to explore data quality issues related to statistical properties of data and get more detailed insights rather than just aggregate numbers.
– Machine learning engineers can preview data issues relevant to their model building and implementation goals that can further aid them in deciding on pre-processing steps or model parameters.
– Domain experts/subject matter experts can analyse the results to take more informed decisions related to corrections/validations as needed.
– Legal or compliance experts can understand and report on aspects like data privacy or use of personal information using metrics that detect bias or use of PI.
– Non-technical users like data owners can understand various quality dimensions of their data in a transparent manner through simple interactions with the tool.

In the following section, we use label purity as an exemplar metric to illustrate the benefit of using effective explanations for remediation of noisy labels in training data using a prototype implementation of our proposed tool.

4 Example: Label Purity

4.1 What Is Label Purity

Label noise is a complex yet important problem associated with annotated data used for training machine learning models [3,9]. Real-world data often contains noisy labels or inconsistent annotations that can have negative consequences when learning a classifier. The source of this noise can be attributed to imperfect evidence, patterns that may be confused with the patterns of interest, perceptual errors or even biological artifacts [3]. The cause is mostly human error resulting from incomplete information, encoding errors, or subjective variability on the right label. Irrespective of the source of noise in data, it is established that label noise decreases the accuracy of predictions, increases model complexity and may require additional training data for building an efficient model.

In our data readiness framework, Label Purity [6] is a metric that quantifies the occurrence of label noise in a given dataset. It does so by detecting label errors in the prediction class and computing a score between 0 and 1 to indicate the magnitude of label noise, where 1 indicates no noise is present in data. The metric also provides suggested labels for each data sample flagged as noisy. Although the suggested labels can be applied automatically, it is preferred and recommended to obtain due review and validation by subject matter experts or data owners. This is because remediation of noisy data involves modification of the original dataset and requires a great degree of confidence and trust in the algorithms' prediction of suggested labels. Putting humans in the loop therefore adds to the robustness and reliability of the process. However this process of reviewing, verification and approval of problematic samples is tedious and time-consuming. The two main reasons making this an expensive exercise are false positives flagged by the algorithm and a lack of effective explanations [2]. We consider *Data-Debugging through Interactive Visual Explanations* as a tool to assist in this task and optimise the productivity of human involvement.

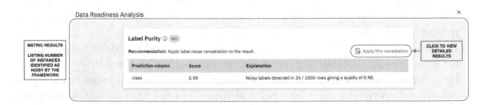

Fig. 3. Label purity results

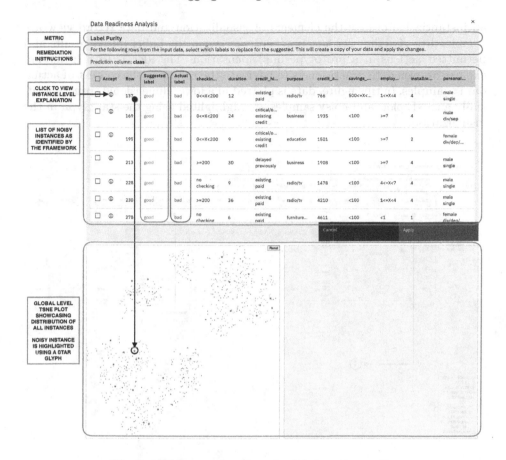

Fig. 4. Building explanations for label purity results

4.2 Explanation

In order to rectify noisy labels identified by the label purity metric, a subject
matter expert is taken into the feedback loop. This is to ensure that corrections
and modifications if any are reliable and trustworthy. The expert not only needs
to inspect the results carefully but also understand them so that appropriate
validations or corrections are applied in an informed manner. This is a largely
manual process and currently there is no easy way for a subject matter expert to
review the noisy samples and their predicted labels. This problem is magnified
when the number of noisy samples in large.

This is an apt use case for our tool where a powerful interface is given to
the human expert for analyzing results, reviewing predictions and applying relevant remediation's. Figure 3 shows the first level of summary results obtained
after running the label purity metric. It shows a score corresponding to amount
of noise detected in data, an explanation of how the score was arrived at and
also a recommendation to run the relevant remediation metric. Figure 4 shows

an instantiation of how the elements of visualization and data record view are combined to give the user a comprehensive view of the Label Purity results and permitted actions in a prototype implementation. The idea is to help human experts or even data practitioners to navigate the results in the context of actual data examples. In this figure, data is plotted in a lower dimensional space, specifically a tSNE plot, and noisy samples are highlighted using the star glyph in the visualization.

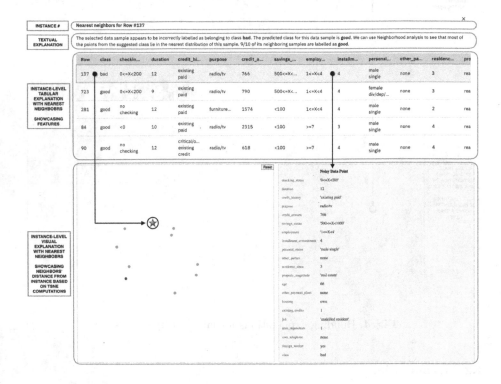

Fig. 5. Instance Level Explanations

There is a one to one mapping between the visualization and data records to present an instance level local explanation to allow inspection of individual data points. Selecting a noisy data point, transforms both the tabular and graphical view to showcase corresponding nearest neighbors(10 in this case) accompanied by textual explanation as shown in Fig. 5. For the selected sample, the textual explanation in natural language is generated using the template *Statement + Evidence*. The statement is a reiteration of the noisy label and the predicted label. The evidence is compiled using neighborhood analysis to explain the reasoning behind the predicted label.

If the global view is insufficient to make the decision, the expert can further investigate the supporting evidence of each detected instance by selecting one

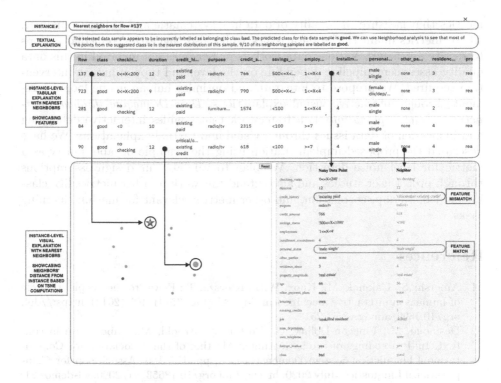

Fig. 6. Neighborhood comparisons

of the neighbors presented in the instance level visualization as shown in Fig. 6. Doing so populates the tabular space accompanying the visualization to highlight the similarities and differences in features of the selected neighbor from the instance being investigated. This helps the decision-making process by understanding which features bring the instance in question closer to entries of other classes and if the suggestion by the framework is acceptable. If convinced, the user can accept the proposed change in the global view for remediation purposes.

Thus, the tool aids the user in navigating the metric results at various levels of granularity using global and local explanations based on user interactions. We anticipate that this combination of visual, textual and data views will reduce the cognitive load for users and the effort required in reviewing. Even though this use case is based on real insights from data practitioners we do aim to get more feedback into the actual design and implementation of the interface through proper user studies in future work.

5   Summary and Future Work

Addressing data quality issues before it enters an ML pipeline allows taking remedial actions to reduce model building efforts and turn-around times. Data Readiness Assessment refers to the various data profiling and quality estimation metrics that measure the quality of data in a systematic and objective manner. These metrics quantify the data issues while also suggesting remedial actions to rectify anomalies. Various data workers or personas are involved in this data quality assessment pipeline depending on the level and type of human intervention required. We propose to integrate and optimize human in the loop assessment through interactive visual explanations. *Data-Debugging through Interactive Visual Explanations* is a framework that enables interactive debugging and inspection of data issues through visual and textual explanations. We have described the proposed implementation using the example of label purity as a metric for label noise detection. We plan to validate our design assumptions through proper user studies and also extend this work to other metrics like class imbalance, overlap, correlation, and other metrics relevant for machine learning tasks.

References

1. Amershi, S., Cakmak, M., Knox, W.B., Kulesza, T.: Power to the people: the role of humans in interactive machine learning. AI Mag. **35**(4), 105 (2014). https://doi.org/10.1609/aimag.v35i4.2513
2. Desmond, M., Finegan-Dollak, C., Boston, J., Arnold, M.: Label noise in context. In: Proceedings of the 58th Annual Meeting of the Association for Computational Linguistics: System Demonstrations, pp. 157–186. Association for Computational Linguistics, July 2020. https://doi.org/10.18653/v1/2020.acl-demos.21. https://www.aclweb.org/anthology/2020.acl-demos.21
3. Frénay, B., Verleysen, M.: Classification in the presence of label noise: a survey. IEEE Trans. Neural Netw. Learn. Syst. **25**(5), 845–869 (2013)
4. Ham, K.: Openrefine (version 2.5) open-source tool for cleaning and transforming data. J. Med. Libr. Assoc. JMLA **101**(3), 233 (2013). http://openrefine.org.free
5. Hohman, F., Srinivasan, A., Drucker, S.M.: TeleGam: combining visualization and verbalization for interpretable machine learning, p. 5 (2019)
6. Jain, A., et al.: Overview and importance of data quality for machine learning tasks, pp. 3561–3562, August 2020. https://doi.org/10.1145/3394486.3406477
7. Kandel, S., Paepcke, A., Hellerstein, J., Heer, J.: Wrangler: interactive visual specification of data transformation scripts. In: Proceedings of the SIGCHI Conference on Human Factors in Computing Systems, pp. 3363–3372 (2011)
8. Mohseni, S., Zarei, N., Ragan, E.D.: A multidisciplinary survey and framework for design and evaluation of explainable AI systems. arXiv:1811.11839 [cs], August 2020
9. Northcutt, C.G., Jiang, L., Chuang, I.L.: Confident learning: estimating uncertainty in dataset labels (2020)
10. Sevastjanova, R., et al.: Going beyond visualization: verbalization as complementary medium to explain machine learning models (2018)
11. Smilkov, D., Thorat, N., Nicholson, C., Reif, E., Viégas, F.B., Wattenberg, M.: Embedding projector: interactive visualization and interpretation of embeddings. arXiv preprint arXiv:1611.05469 (2016)
12. Spinner, T., Schlegel, U., Schafer, H., El-Assady, M.: Explainer: a visual analytics framework for interactive and explainable machine learning. IEEE Trans. Vis. Comput. Graph. 1 (2019). https://doi.org/10.1109/TVCG.2019.2934629

Data Augmentation for Fairness in Personal Knowledge Base Population

Lingraj S. Vannur[1], Balaji Ganesan[2(✉)], Lokesh Nagalapatti[2], Hima Patel[2], and M. N. Tippeswamy[1]

[1] Nitte Meenakshi Institute of Technology, Bengaluru, India
`thippeswamy.mn@nmit.ac.in`
[2] IBM Research, Bengaluru, India
{`bganesa1,lokesn21,himapatel`}`@in.ibm.com`

Abstract. Cold start knowledge base population (KBP) is the problem of populating a knowledge base from unstructured documents. While neural networks have led to improvements in the different tasks that are part of KBP, the overall F1 of the end-to-end system remains quite low. This problem is more acute in personal knowledge bases, which present additional challenges with regard to data protection, fairness and privacy. In this work, we use data augmentation to populate a more complete personal knowledge base from the TACRED dataset. We then use explainability techniques and representative set sampling to show that the augmented knowledge base is more fair and diverse as well.

Keywords: Data augmentation · Fairness · Personal knowledge base

1 Introduction

The NIST TAC Knowledge Base Population (KBP) challenges introduced the cold start knowledge base population problem. The problem involves several tasks that go into populating a knowledge base from unstructured documents. Figure 1 shows a typical pipeline that includes entity recognition, entity classification, entity resolution, relation extraction, and slot filling.

As [20] showed while introducing the TACRED dataset, this problem remains largely unsolved, with the end-to-end system achieving only 26.7 F1, which is not sufficient for real world applications. This is even so when tasks like relation extraction have achieved 74.8 F1 [6].

There have been efforts to understand the low overall F1 in the TACRED dataset. [1] investigated the harder samples in the TACRED dataset and concluded that some of the labels need to be re-labeled. Our work examines the TACRED dataset from a different perspective. We populate a personal knowledge base from the TACRED dataset and investigate the diversity of the protected attributes like gender, ethnicity, location and religion.

Electronic supplementary material The online version of this chapter (https://doi.org/10.1007/978-3-030-75015-2_15) contains supplementary material, which is available to authorized users.

M. Gupta and G. Ramakrishnan (Eds.): PAKDD 2021 Workshops, LNAI 12705, pp. 143–152, 2021.
https://doi.org/10.1007/978-3-030-75015-2_15

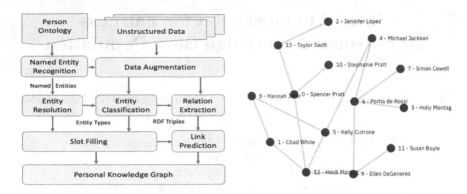

(a) Personal Knowledge Base Population   (b) Personal Knowledge Base

Fig. 1. Our goal is to populate a Personal Knowledge Base from TACRED dataset

Constructing a property graph of people like in Fig. 1b, from unstructured documents like emails in an organization, news articles, personal conversations etc. is a significantly harder problem than populating general purpose world knowledge bases. This is because, without densely populated node attributes, tasks like entity resolution and link prediction become harder.

While populating a personal knowledge base, we have to strive to avoid bias in the training data on gender, age, ethnicity, location, religion, sexual orientation among other attributes. While there are methods available to detect bias in models, eliminating bias in the training data helps business intelligence and other analytics applications as well. We propose data augmentation as the solution to these different challenges.

Data augmentation is the process of increasing the diversity in the training data without necessarily having to acquire more data. In this work, we show how the popular TACRED Dataset can be made more diverse by addressing the low recall in extracting protected attributes. Our motivation is to have more features to train Graph Neural Networks, while also increasing the overall diversity of the populated Personal Knowledge Base.

After augmenting the data used for knowledge base population, we need to be able to show that the resulting graph is more diverse and fair. We use explainability techniques to observe the effect of augmentation on the Link Prediction task. This is to ensure that the predictions are not unduly dependent on protected attributes. Another technique to observe the improvement in diversity is to sample the data and visually observe the samples. We propose using representative set sampling for this validation.

2   Related Work

Knowledge Base Population (KBP) has been a fairly well researched problem. The KBP Track at TAC [11] that were held between 2010 and 2017 led to significant advances in KBP. Recently [14] have introduced a benchmark dataset for the KBP problem. In this work though we focus more on the Personal Data in the Knowledge Base, given our goal to make the KBP process fair.

[3] introduced the concept of Personal Knowledge Graphs. In this paper, we describe a system to extract personal knowledge graph from the TACRED relation extraction dataset. We treat this as an acceptable proxy for real world enterprise documents like emails, internal wiki pages, organization charts which cannot be used for research because of privacy reasons.

[4] introduced a AI Fairness toolkit called AIF360 which provides a number of algorithms for detecting and mitigating bias against protected attributes like gender, age, ethnicity, location, sexual orientation and few others. [9] proposed a method using counter factual data to improve fairness.

[17] introduced the idea of Anchors as Explanations, which builds upon their earlier LIME solution [16]. The key idea here is to show only few important features (anchors) rather than showing the pros and cons of several or all features.

[10] introduced GraphLIME, which as the name indicates is a version of LIME for Graph Neural Networks. They compared their solution with GNN Explainer [18] which follows a subgraph approach to explain predicted links. [19] proposed a model level explanation solution using reinforcement learning. In this work, we have used Anchors along with LIME [16] and SHAP [13] for the fairness analysis. For other kinds of explanations, a graph based explainability solution will be more suitable.

3   Data Augmentation

Any system that assigns a label to a span of text can be called an annotator. For data augmentation task, we have used SystemT [5], OpenIE [2] and custom regex based annotators. We'll now briefly explain annotations using SystemT and refer to [2] for details on the OpenIE annotations.

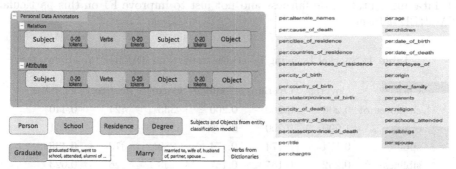

(a) Personal Data Annotators based on SystemT   (b) TACRED Person relations

Fig. 2. Data augmentation

146   L. S. Vannur et al.

As shown in Fig. 2a we can define generic patterns to identify *Subject, Object* and *Predicates*. We can also annotate individual attributes like name of a person, school, place of residence, academic degree by populating dictionaries for each of those attributes. This approach suffices in our particular case, because we are trying to augment the attributes and relations that are already present in the TACRED dataset. If we were to annotate from scratch, we may have to consider other sophisticated approaches. SystemT does not take the context of the entity mentions while assigning labels and hence is more suited to coarse grained entity types rather than finer types. However, labels for name, email address, location, website do not suffer much from this lack of context.

Once we have additional annotators, we need to be able to evaluate them and select only the correct labels or the ones presumed to be correctly labelled. We have used the Snorkel [15] system which uses data programming approach and requires only a small amount of manual annotations.

As can be seen in Table 1, our annotators (combining SystemT, OpenIE and custom labeling functions) perform reasonably well in annotating features but less well in annotating new relations.

We compared annotations from our labeling functions with the original annotations in TACRED. We manually verified around 1000 attributes and relations in the augmented dataset. In Table 2, as an example, we show the detailed performance of each set of labeling functions on one attribute (org) and one relation (employee_of). The original TACRED annotations continue to agree with our manual observations to a large extent. We discuss the overall performance of the KBP evaluation in Sect. 4.

As shown in Fig. 3, we were able to improve the number of attributes and relations substantially with a relatively modest effort of writing labelling functions. We annotated gender and ideology which were not in the original annotations for the specific requirements of this work to measure fairness.

We note here that using language models, and neural models described in [7], we can further improve annotations. However, we now move on to analysing the effect of these annotations, since our focus here is to demonstrate the effect of data augmentation on fairness and not just to improve F1 on this particular TACRED dataset.

Table 1. Snorkel Labeling functions analysis on TACRED relations

Class	Coverage	Overlaps	Conflicts	Correct	Incorrect	EmpAcc
employee_of	0.200	0.200	0.200	333	424	0.440
org	1.000	0.200	0.200	2501	1280	0.661
spouse	0.174	0.174	0.174	289	370	0.438
siblings	0.252	0.157	0.157	43	26	0.623
wife	0.274	0.274	0.157	46	29	0.613

Table 2. Comparison of annotations by different labeling functions

Class	Annotator	Coverage	Overlap	Conflict	Correct	Incorrect	EmpAcc
org	tacred	0.219	0.169	0.100	271	58	0.824
org	custom	0.200	0.199	0.093	118	182	0.393
org	openie	0.031	0.025	0.019	9	37	0.196
org	systemt	0.200	0.199	0.093	114	186	0.380
employee	tacred	0.220	0.143	0.066	287	43	0.870
employee	custom	0.215	0.199	0.087	101	222	0.313
employee	openie	0.091	0.081	0.076	31	106	0.226
employee	systemt	0.215	0.198	0.104	68	255	0.211

4 KBP Evaluation

In this section, use the evaluation method in the TAC KBP 2015 cold start slot filling task [8]. This method involves evaluating entities called hop-0 and hop-1. These are manually selected by random search for people and their relations in the corpus data. Hop-0 generally involves the person to person relation and hop-1 consists of person to attributes relations. The evaluation is done by comparing the precision and recall in the extracted triples with the manually annotated ground truth. Figure 3 shows the improvement in extracting select attributes and relations.

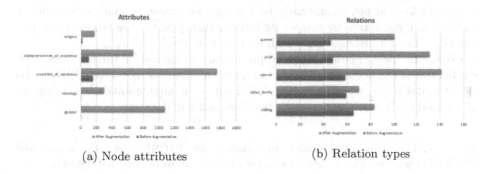

(a) Node attributes   (b) Relation types

Fig. 3. A sample of the augmentation on attributes and relations

As shown in Table 3 recall of our augmented dataset is 0.518 is better than that of TACRED baseline recall. In particular, we observe that the recall of protected attributes before augmentation is only around 0.095 whereas the recall of protected attributes after augmentation is around 0.99.

Table 3. Comparison of baseline and augmented annotations using our manual ground truth on TACRED dataset

Metric	TACRED			TACRED+		
	Precision	Recall	F1	Precision	Recall	F1
hop0	0.449	0.307	0.365	0.643	0.553	0.595
hop1	0.429	0.429	0.256	0.615	0.482	0.541
hop-all	0.439	0.368	0.311	0.629	0.518	0.568

5   Fairness Analysis

For fairness analysis, we train a random forest model to predict whether a person will have more than one link. The assumption here is that a singleton node or being a peripheral node is less desirable and hence we check if this happens to minority groups. This trivial classifier perhaps suffices to demonstrate if the prediction is based on protected variables. More complex predictions can be used to analyze more sensitive predictions in real world applications.

As shown in Table 4, Position Aware Graph Neural Network performs better than Graph Convolutional Networks in all three of our experiments. We first evaluate Link Prediction performance on the original TACRED dataset and then repeat on the augmented TACRED* dataset.

We conduct experiments keeping the minimum size of subgraphs (connected components) at 5 and 10. Given the sparse nature of the original graph, there were no subgraphs of size more than 5 in the TACRED dataset. While the data augmentation did not significantly improve performance on small subgraphs, it enabled link prediction by producing larger subgraphs. On the larger subgraphs the GCN has 0.57 ROC AUC which is more reasonable.

While the performance of the graph neural models can be further improved, we believe our experiments show that popular datasets like TACRED can be made suitable for training GNNs. Without such augmentation, training an attributed GNN made not even be feasible on many datasets.

Table 4. Link Prediction models performance on TACRED and the augmented TACRED Datasets. There were no subgraphs of size 10 in the original dataset.

Dataset	Model	Min subgraph	size 5	Min subgraph	size 10
		ROC AUC	Std. Dev.	ROC AUC	Std. Dev.
TACRED	GCN	0.25	0.25	–	–
	P-GNN	0.50	0.00	–	–
TACRED+	GCN	0.23	0.14	0.50	0.11
	P-GNN	0.46	0.037	**0.57**	0.01

Next we analyze the features. We use LIME [16] to rank the features that influence the predictions of the random forest classifier which is trained on the output of the PGNN model. This post-hoc explainability using an interpretable model remains popular in the literature, since self-explainable models are still not mature.

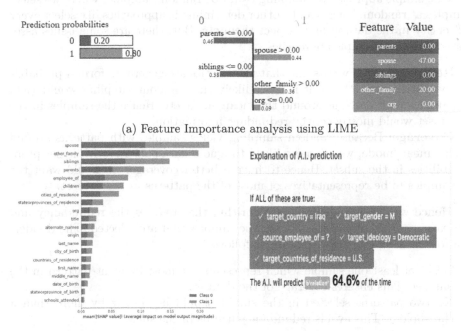

(a) Feature Importance analysis using LIME

(b) Feature Importance analysis using SHAP values
(c) Feature Importance analysis using Anchors only on features

Fig. 4. Fairness analysis using LIME, SHAP and Anchors

In Fig. 4a, blue bars represent the features that are not contributing to the prediction and the orange bars show the features that are contributing to the prediction. We observe that features such as organization, family relations are main attribute to predict whether a new person added to the Knowledge Base is going to be linked to any other person in the graph.

While LIME is used to analyze individual predictions, use SHAP [13] to analyze the model level interpretations. As shown in Fig. 4b, predicting whether a node will be a singleton or leaf node, seems to be predominantly dependent on edges (as expected) and features like city, org and less on sensitive attributes. However, using the Anchors [17] explainability method, we see some examples where gender is an important feature. We believe data augmentation followed by human evaluation using the above explainability techniques, and iterating on this could be the best approach to ensure fairness in personal knowledge graph population.

6   Representative Set Sampling

Now we present our second approach to measure the diversity of the augmented dataset. We propose that sampling the dataset before and after augmentation could show the increase in diversity, if any.

One simple approach to sampling could be random sampling where we select samples at random. There can be other deterministic approaches like select every k^{th} entity in the dataset and inspect them etc. But, there are some issues associated with such simple approaches.

- **Redundancy:** If we assume that the entities are generated form a probability distribution P, then it is more likely that random sampling would yield samples with patterns around the mean/mode etc. Hence the samples in the subset would mostly convey redundant information.
- **Coverage:** Because random sampling yields samples with patterns around the mean/mode, we are more likely to ignore data patterns with lesser probabilities in the subset. Hence to have a better coverage, we would want the samples to be representatives of most of the patterns in the dataset.

Hence we develop a sampling algorithm that reduces the redundancy and increase coverage of patterns among the samples that are selected. The desiderata for the sampling algorithm are as follows:

- Select at least one sample which represents the most frequent pattern in the dataset. This ensures coverage of the dataset.
- No two patterns selected in the subset should be similar by more than a threshold θ. This avoids redundant patterns in the subset.

Our sampling algorithm is inspired from matrix sketching [12], which is a linear time algorithm to find the most frequent patterns in the dataset. More formally, given a matrix $A \in \mathcal{R}^{n \times m}$, the algorithm finds a smaller sketch matrix $B \in \mathcal{R}^{l \times m}$ such that $l << n$ and $A^T A \approx B^T B$. Here we can observe that the matrix B tries to capture most of the variance in A. In other words, each row of B represents a frequent direction in A and also because B is obtained by performing SVD on rows of A, each row of B is orthogonal to other rows of B.

Our intuition is that, once we get the frequent directions of A, we can easily select data points along that direction and thereby select samples representing the frequent patterns in the dataset. The sampling algorithm expects the input to be in numerical form only. We convert each categorical attribute to one hot embedding and normalize each numerical column to be between $[0, 1]$ and feed it as input to the algorithm. We drop other text attributes. Hence, input to the sampling algorithm is a matrix A that is scaled for numerical attributes and one-hot embedded for categorical attributes respectively.

The results of the representative set sampling are shown in Fig. 5a. The algorithm selected a total of 26 and 222 data points respectively from the baseline TACRED dataset and our augmented dataset.

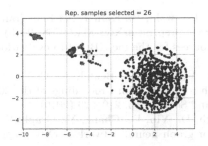 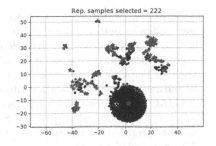

(a) Sampling before augmentation   (b) Sampling after augmentation

Fig. 5. TSNE projections of the dataset with blue dots denoting the points in the dataset and red dots denoting the representative samples. The more diverse the dataset, more samples will be needed to adequately represent the dataset.

7 Conclusion

Using different labelling functions and the Snorkel data programming approach, we showed how data augmentation improves the overall diversity of a personal knowledge base populated from the TACRED dataset. We then performed a detailed fairness analysis and representative set sampling on the augmented dataset. Our work seems to show that data augmentation can help train models with reasonable confidence that they are not biased against minority groups.

Acknowledgement. This work was done as part of the Global Remote Mentoring initiative of IBM University Relations to promote undergraduate student research. We thank Kalapriya Kannan, Dinesh Garg, Poornima Iyengar, Kranti Athalye, and Nitte Meenakshi Institute of Technology for their support.

References

1. Alt, C., Gabryszak, A., Hennig, L.: TACRED revisited: a thorough evaluation of the TACRED relation extraction task. arXiv preprint arXiv:2004.14855 (2020)
2. Angeli, G., et al.: Bootstrapped self training for knowledge base population. In: TAC (2015)
3. Balog, K., Kenter, T.: Personal knowledge graphs: a research agenda. In: Proceedings of the 2019 ACM SIGIR International Conference on Theory of Information Retrieval, pp. 217–220 (2019)
4. Bellamy, R.K., et al.: AI fairness 360: an extensible toolkit for detecting, understanding, and mitigating unwanted algorithmic bias. arXiv preprint arXiv:1810.01943 (2018)
5. Chiticariu, L., Krishnamurthy, R., Li, Y., Raghavan, S., Reiss, F.R., Vaithyanathan, S.: SystemT: an algebraic approach to declarative information extraction. In: Proceedings of the 48th Annual Meeting of the Association for Computational Linguistics (2010)
6. Cohen, A.D., Rosenman, S., Goldberg, Y.: Relation extraction as two-way span-prediction. arXiv preprint arXiv:2010.04829 (2020)

7. Dasgupta, R., Ganesan, B., Kannan, A., Reinwald, B., Kumar, A.: Fine grained classification of personal data entities. arXiv preprint arXiv:1811.09368 (2018)
8. Ellis, J., et al.: Overview of linguistic resources for the tac KBP 2015 evaluations: methodologies and results. In: TAC (2015)
9. Garg, S., Perot, V., Limtiaco, N., Taly, A., Chi, E.H., Beutel, A.: Counterfactual fairness in text classification through robustness. In: Proceedings of the 2019 AAAI/ACM Conference on AI, Ethics, and Society, pp. 219–226 (2019)
10. Huang, Q., Yamada, M., Tian, Y., Singh, D., Yin, D., Chang, Y.: GraphLIME: local interpretable model explanations for graph neural networks. arXiv preprint arXiv:2001.06216 (2020)
11. Ji, H., Grishman, R., Dang, H.T., Griffitt, K., Ellis, J.: Overview of the tac 2010 knowledge base population track. In: Third Text Analysis Conference (TAC 2010), vol. 3, p. 3 (2010)
12. Liberty, E.: Simple and deterministic matrix sketching (2012)
13. Lundberg, S.M., Lee, S.I.: A unified approach to interpreting model predictions. In: Advances in Neural Information Processing Systems, pp. 4765–4774 (2017)
14. Mesquita, F., Cannaviccio, M., Schmidek, J., Mirza, P., Barbosa, D.: KnowledgeNet: a benchmark dataset for knowledge base population. In: Proceedings of the 2019 Conference on Empirical Methods in Natural Language Processing and the 9th International Joint Conference on Natural Language Processing (EMNLP-IJCNLP), pp. 749–758 (2019)
15. Ratner, A., Bach, S.H., Ehrenberg, H., Fries, J., Wu, S., Ré, C.: Snorkel: rapid training data creation with weak supervision. VLDB J. 1–22 (2019)
16. Ribeiro, M.T., Singh, S., Guestrin, C.: "Why should i trust you?" Explaining the predictions of any classifier. In: Proceedings of the 22nd ACM SIGKDD International Conference on Knowledge Discovery and Data Mining, pp. 1135–1144 (2016)
17. Ribeiro, M.T., Singh, S., Guestrin, C.: Anchors: high-precision model-agnostic explanations. In: AAAI, vol. 18, pp. 1527–1535 (2018)
18. Ying, Z., Bourgeois, D., You, J., Zitnik, M., Leskovec, J.: GNNExplainer: generating explanations for graph neural networks. In: Advances in Neural Information Processing Systems, pp. 9244–9255 (2019)
19. Yuan, H., Tang, J., Hu, X., Ji, S.: XGNN: towards model-level explanations of graph neural networks. arXiv preprint arXiv:2006.02587 (2020)
20. Zhang, Y., Zhong, V., Chen, D., Angeli, G., Manning, C.D.: Position-aware attention and supervised data improve slot filling. In: Proceedings of the 2017 Conference on Empirical Methods in Natural Language Processing, pp. 35–45 (2017)

The First International Workshop on Artificial Intelligence for Enterprise Process Transformation (AI4EPT 2021)

ROC Bot: Towards Designing Virtual Command Centre for Energy Management

Rishi Tiwari, Mohammed Afaque, Amit Sangroya$^{(\boxtimes)}$, and Mrinal Rawat

Tata Consultancy Services, Mumbai, India
{t.Rishi,Afaque.m,amit.sangroya,rawat.mrinal}@tcs.com

Abstract. Domains such as energy management rely heavily on dashboards and other related interfaces to manage the infrastructure and resources. The users of this domain use dashboards to manage the data and extensively perform periodic analysis to save energy and cost. Creating multiple dashboards for visualization of data is not user-friendly from a design perspective. This motivates the need of a single interface through which users can do data exploration, visualization and summarizing. Combining this with features such as anomaly detection can identify various issues and assist in day to day monitoring of an energy management center.

In this paper, we present ROC (Resource Optimization Center) Bot, a novel data exploration tool with a natural language interface. ROC Bot leverages recent advances in deep models to make query understanding more robust in the following ways: First, ROC Bot uses a deep model to translate natural language statements to SQL, making the translation process more robust to paraphrasing and other linguistic variations. Second, to support the users in automatically summarizing data, ROC Bot provides a machine learning model that helps in writing natural looking summaries in any given tabular data.

Keywords: Automation · Deep learning · Conversational systems

1 Introduction

Huge amount of data is stored and is being accessed on a daily basis in large corporations. This emerging chunk of information has lead to solutions which have taken up the market space and have gained a wide popularity. Organisations create bulky and time consuming interfaces to manage the data, processes and other activities. These interfaces provide assistance to the user in accomplishing their tasks but often require time to be adapted.

Providing users with dashboards pose challenges based on the familiarity and the user experience leveraged. This causes users to follow a lot of procedures in order to accomplish minor tasks as well. Dashboards provide the users with an interface to interact with the databases by using some programmed procedures.

© Springer Nature Switzerland AG 2021
M. Gupta and G. Ramakrishnan (Eds.): PAKDD 2021 Workshops, LNAI 12705, pp. 155–167, 2021.
https://doi.org/10.1007/978-3-030-75015-2_16

The dashboards need to be programmed again to achieve new features. There are multiple activities in a corporate environment which are to be done manually and manuals are prepared to accomplish them. All the activities mentioned previously provide challenges and consume more time than necessary. To overcome this problem, we present our solution that eliminates the need of multiple screens and dashboards and takes care of the necessary activities with minimal intervention.

We are providing a smoother and easier way of minimizing the use of multiple screens. We have implemented a single conversational interface which takes care of all these challenges and also provides an ease to the end user by eliminating the need of dashboards which consume more time and effort. To connect and access data from a database we have implemented a pipeline of deep learning models which convert the natural language sentences to database queries. An AI engine is responsible for the data analysis, optimization and data summarizing. This engine handles activities and prompts the users for input in real time. Since the interface is conversational intervention is done only when necessary.

Contributions: In this paper, we introduce ROC-Bot, a conversational interface that provides a robust and easy-to-use natural language (NL) interface with the purpose of improving and enhancing the expressiveness and flexibility of human-data-interaction. Different from existing approaches, ROC-Bot leverages deep neural network models as the core of its natural language interface system. In the following, we outline the key features of ROC-Bot:

Robust NL Query Mapping: We propose a novel NL to database query mapping framework based on a sequence-to-sequence recurrent neural network model that can perform efficient queries on knowledge graphs. Our notion of model robustness is defined as the effectiveness of the ML model to map linguistically varying utterances to finite predefined relational database operations. For example, there are numerous ways in which a query can be asked in a natural language, such as "show all buildings where energy consumption is below average; show below average energy consumption buildings".

A key challenge hereby is to curate a comprehensive training set for the model. While existing approaches for machine translation require a manually annotated training set, we implement a novel synthetic generation approach that uses only the database schema with minimal annotation as input and generates a large collection of pairs of natural language queries and their corresponding database query statements.

Auto-completion: We provide real-time auto completion and query suggestions to help users who may be unfamiliar with the database schema or the supported query features. This helps to improve translation accuracy by leading the user towards less ambiguous queries. Consider a scenario in which a user is exploring an energy management database and starting to type "show me the energy" – at this point, the system suggests possible completions such as consumption, utilization, or wastage to make the user aware of the different options she has, given the specific database context. At the core of the auto-completion feature is a language model based on the same sequence-to-sequence architecture and trained on the same synthetic training set as the query translator

Auto-summarization: The problem of generating textual summaries from structured tabular data is finding prominent applications in automating business procedures that currently rely on manual report/summary generation. Prior approaches to summary generation have identified two key sub-tasks i) structured content selection and ii) summary generation. A glaring limitation of prior solutions is that all numeric data in the tables is encoded via text embeddings. However, textual embeddings cannot reliably encode information about numeric concepts and relationships. Moreover, since every numeric value has a unique embedding, generalizing to unseen numeric values in test data becomes challenging. In this paper, we propose to address this problem by incorporating numerical information for different tabular fields and allowing the system to learn correlations between the rank of a numeric value and the probability of its inclusion in the resulting summary. We propose a novel architecture that employs a combination of an LSTM and a pointer network for content selection and an LSTM based encoder decoder for summary generation.

2   Related Work

There are a lot of conversational AI based personal assistant tools already in market for the end users. Some of these tools are designed for open-domain conversations e.g. Cortana, Siri, Alexa and Google Now [2,13]. On the other hand there are also tools where focus is on supporting interactive data assistance e.g. Microsoft Power BI [8]. However, there are still limitations of the existing solutions in terms of natural language based conversational capabilities while supporting data assistance. There are still various open issues in terms of semantic understanding and context-awareness of users' queries. Most important task for building a NL based data assistant is to generate structural query language (SQL) queries from natural language. Answering a natural language question about a database table requires modeling complex interactions between the columns of the table and the question. Jedeja et al. presents four different perspectives namely user experience, information retrieval, linguistic and artificial intelligence for the evaluation of conversational AI systems [4].

The study of translating natural language into SQL queries has a long history. Popescu et al. introduce *Precise NLI*, a semantic parsing based theoretical framework for building reliable natural language interfaces [11]. This approach relies on high quality grammar and is not suitable for tasks that require generalization to new schema. Recent works consider deep learning as the main technique. There are many recent works that tackle the problem of building natural language interfaces to relational databases using deep learning [1,6,9,16–20]. Zhong et al. [20] propose *Seq2SQL* approach that uses reinforcement learning to break down NL semantic parsing task to several sub-modules or sub-SQL incorporating execution rewards. Yavuz et al. introduce *DialSQL*, a dialogue based structured query generation framework that leverages human intelligence to boost the performance of existing algorithms via user interaction. The flexibility of our approach enables us to easily apply sketches to a new domain. Our framework also does not require large corpus of NL sentences as training input.

3   ROC Bot: Virtual Command Centre for Energy Management

ROC Bot is designed as a set of intelligent suite of AI tools designed to help business analysts in getting insights about data in a user friendly way. Instead of meandering through the database for a small detail, ROC Bot provides an interface where business analysts just need to type their query in Natural Language (English) and system will present the results. ROC Bot supports context understanding by precisely capturing user's intent and responding appropriately.

More often, business analysts have a range of queries that they perform to get data insights. Sometimes they just need a one line answer for their questions such as *"What is phone number of Julia"*. Sometimes it might be some aggregation query e.g. *"How many employees are hired in 2018"* and in some cases they might be interested in visualizing the information in the form of charts e.g. *"Show me the acceptance rate of EMNLP over last ten years"*. Therefore, ROC Bot is designed to handle all these variations of NL queries. It supports a conversational interface that business analysts can use and get their information in fraction of seconds. It makes quick, efficient, and easy interactions with data. Depending on the nature of query, ROC Bot can also respond with appropriate visualized interpretation of the data in the form of Pie Charts, Graphs etc.

We propose an architecture where a database query is formed from a natural language sentence with the help of an intermediate form i.e. query sketch. Our system comprises of two parts: a) Mechanism to generate an intermediate form (**Query Sketch**) given a NL sentence and b) Approach of transforming a sketch to database query.

3.1   Deep Learning Based Framework for Query Sketch Generation

In order to generate the query sketch, we have a pipeline of multiple sequence tagging deep neural networks. Our architecture consists of a bidirectional LSTM network along with a CRF (conditional random field) output layer. In our architecture framework, the sequence of word embedding is given as input to a bidirectional LSTM. Instead of using the softmax output from this layer, we use a CRF Layer yielding the final predictions for every word. We use ELMO embedding that are computed on top of two-layer bidirectional language models with character convolutions as a linear function of the internal network states [10]. The character-level embedding have been found useful for specific tasks and to handle the out-of-vocabulary problem. The character-level representation is then concatenated with a word-level representation and feed into the Bidirectional LSTM as input. The intent is to identify the parts in the sentence which are relevant. We annotate a NL query as follows. { *What is the employee id of*

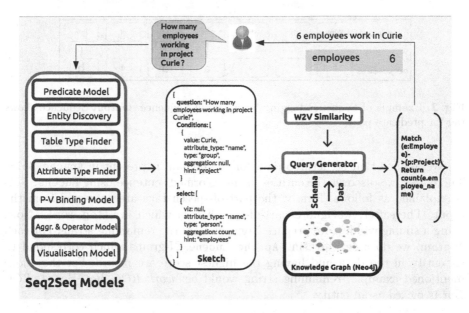

Fig. 1. Models for NL to Query generation

John} is annotated as {0 0 0 A A 0 B} (See Figs. 2 and 3 for reference schema and example sequence respectively) (Fig. 1).

This example shows that there are two concepts *employee id* and *John* marked as A and B respectively. The same token is used for the concepts which consist of more than one word e.g. *employee id* consists of two words, so we mark them with token 'A'. We follow this representation in our models. This approach makes our system database agnostic and even with comparatively less amount of training data we are able to extract the information out of the sentence. Since models are dependent on each other, we will explain each of them with the help of an example: *How many employees work in project ROC Bot?*

Predicate Finder Model. This model finds the target concepts (**predicates**) *P* from the NL sentence. In case of database query language, predicate refers to the SELECT part of the query. Once predicates are identified, it becomes easier to extract entities from the remaining sentence. The input to the model is a vector form representation of NL sentence. For example, a natural language sentence, {*How many employees work in project ROC Bot*} is annotated as {0 0 A 0 0 B 0}. In this example *employees* and *project* are predicates.

Entity Discovery Model. This model aims to find the **values/entities** in the sentence. The output from predicate model is taken and reformed as: {*How many <predicate> work in <predicate> ROC Bot*}. The predicates are replaced with *<predicate>* token.

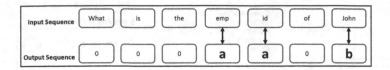

Fig. 2. Example of Sequence tagging for an Input Sequence. Output Sequence shows tag for predicate part.

We assume that structured data for the domain is present in `Apache-Solr`. Thereafter, we discover the entities in the reformed sentence using `Lucene`. This is explained as follows. Firstly, the part-of-speech tags are extracted from the input. Thereafter, we ignore part-of-speech tags which are stop words. Now, using a sliding window we prepare N-grams from the remaining string. For each N-gram, we do a search using `Apache Lucene`. N-grams which correspond to an entity in the data and having the highest score are picked. For the above mentioned example, remaining string would be: *work ROC Bot*. Hence, *ROC Bot* is picked as an entity.

Type Level Finder. This model identifies the **type of concepts** (predicates and values) at the node or table level. For example, *employee id* belongs to *Employee* node and *Employee* is a *PERSON*, so the type of *employee id* is *PERSON*.

If a concept is present in more than one table, **type** information helps in the process of disambiguation. For example, consider following sentences:

- What is the employee id of Washington?
- List all the stores in Washington.

Here, in first example *Washington* refers to the name of a person, whereas in second example it is the name of a location. In such cases, this model is useful to disambiguate that in first example the node level type is `PERSON` and in second example it is `LOCATION`. This helps in making the overall framework database agnostic. In this model, all the entities in input are marked with tag <*value*>. For example, natural language sentence, {*How many employees work in project* <*value*>} is annotated as {0 0 person 0 0 project project}.

Attribute Level Type Finder. This model identifies the **attribute** type of concepts (predicates and values). For example, let's take two sentences:

- What is the employee id of May?
- List all employees hired in May?

In these examples, *May* is present in the same table *Employee*, but refers to different attributes. The attribute type information model here can easily distinguish based on the nature of query that in first example, *May* is the name

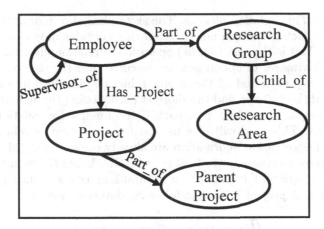

Fig. 3. Schema example of enterprise data

of a person and in second example it is *date*. The attribute type information in combination with node type information helps in increasing the accuracy. In this model, all the entities in input are marked with tag <*value*>. For example, {*How many employees work in project* <*value*>} is annotated as: {0 0 name 0 0 name name}.

Predicate Value Finder. In some queries **predicate value bindings** are already present. For example, let's take two examples:

- How many employees work in ROC Bot?
- How many employees work in project ROC Bot?

In first example, there is no information about the concept *ROC Bot*, but the second example describes that *ROC Bot* is some project. This model binds the predicate, in this example *project* to the value or entity *ROC Bot*. We replace the predicates with tag <*predicate*> and entities with tag <*value*>. {*How many* <*predicate*> *work in* <*predicate*> <value>} is annotated as: {0 0 0 0 0 0A A}. The tokens at index 5 and 6 are binded as [*Project*] ?? [*ROC Bot*]. "??" slot will be filled with an operator that we illustrate in next model.

Generating Database Query from Sketch. The process of generating the query is independent of underlying database i.e. the same approach can be used for generating queries across databases. We demonstrate this concept using two popular relational database query languages: SQL (structured query language) and CQL (cipher query language).

Provided with a sketch **S**, we generate a database query **Q**. As explained earlier, sketch contains information about all the required predicates **L**, i.e. the list of attributes/columns that need to be extracted from Table **T**. This component will give a new relation that only contains the columns in **L**. This operation is projection (Π) in relation algebra. The sketch has also information about all conditions φ which gives us a new relation that contains only those rows satisfying φ in $\mathbf{T}$. This is the selection (σ) operation of relation algebra. To generate the query following set of operations are executed over the sketch:

The next step is to find all the unique tables (U) involved in a given sketch $\mathbf{S}$. If the length(U) > 1, we find the shortest path passing through all the unique tables. We assume that there is a path if the foreign key relationship exists between tables. This step will give us join ($\bowtie$) operation of relation algebra. Operator and aggregation information are already present in the sketch S. Now, combining these, we compute the final relation ($\mathbf{R}$). Using $\mathbf{R}$, we finally compute the DB specific query. For example, a natural language sentence { *"How many employees in each project"*} leads to following database query:

$$\Pi_{count(employee\_id),g(project\_name)}$$
$$(Employee_{pid\_fk} \bowtie_{pid} Project)$$

3.2   Analytics, Automation and Anomaly Detection

Our solution also provides the ability to perform analytics as well as plugged in automated tasks to reduce effort, time and cost involved. The organisation required the real time analysis of data to optimise the energy units and cost. Our solution made it possible with ease as the automated activities are being done by the solution and the end users are prompted on completion or requirement of any input. The analytics and automation part is directly connected with the common conversational interface and communicates using the same with users.

The major analysis was with the information which is changing over time. There are energy parameters such as equipment units which have to be monitored on the basis of the occupancy of people in a particular accommodation area. This monitoring when done with conventional methods require some screen to be viewed constantly and then to take proactive measures in order to optimise power. Using the proposed framework, this issue was eliminated and users can just log into the main interface and the system automatically analyses the parameters in real time. This brought a significant saving in terms of power and cost.

Anomaly detection is a basic requirement in the modern era of analytical solutions. There is an influx of information which pertains to a definitive pattern. In the case of an anomaly it is necessary that it needs to be corrected, fixed and be alerted to the user. Streams of power units, occupancy information, running patterns and a lot more are existent in the energy management domain and having an anomaly detector churns out a great deal of benefit for the user.

3.3   Summarization

A classic problem in natural-language generation (NLG) involves taking structured data, such as a table, as input, and producing text that fluently and

Table 1. Table and Summary from HR domain

Employee ID	Employee name	Age	Gender	Salary
1001	Kartik	20	M	10000
1002	Michael	25	M	20000
1003	Peter	30	M	30000
1004	Raj	35	F	40000

With order model Summary: There are total 4 employees. Average age of employee is 27.5. There are 3 males and 1 female. Highest salary is 40000. Lowest salary is 10000. Average salary is 25000.

Without order model Summary: There are total 4 employees. Average age of employee is 27.5. There are 3 males and 1 female. Highest salary is 40000. Lowest salary is 10000. Average salary is 25000.

succinctly narrates this data as output [5,7,12,15]. The solution to this problem comprises of two important tasks: 1) selecting a subset of the data to report, also called content selection and 2) generation of a natural language summary using this subset. We propose an approach that captures the numeric information available in the table along with the entity information. Consequently, we show that our model generalizes to new entities better than other state of the art models.

Step 1: Record Selection & Ordering Encoding Numeric Entities. Previous works encode all numeric data in the tables via text embeddings. However, textual embeddings cannot reliably encode information about numeric concepts and relationships. Moreover, since every numeric value has a unique embedding, generalizing to unseen numeric values in test data becomes challenging. To overcome this we introduce ranks for all the entities based on their points. This ensures that the *"points"* information is encoded. This also helps in handling unseen entities in the test data. A new unseen entity in the test data has a representation according the numeric information such as *points, rebounds, assists* etc. Along with entity information, this information is important and plays a role in deciding if a particular fact has to be there in generated summary or not, unlike in prior models. For example, a player scoring maximum points will always be there in the generated summaries. This model ensures that player scoring maximum points will always have a place in the generated summary.

Record Embeddings. Building on prior work, we integrate rank information and use an embedding layer for every feature to obtain a vector representation. The final representation is a concatenation of individual feature embeddings.

$$r_j = (\mathbf{W}_r[r_{j,1}; r_{j,2}; r_{j,3}; r_{j,4}; r_{j,5}]) \tag{1}$$

where $[;]$ indicates vector concatenation, $\mathbf{W}_r \in \mathbb{R}^{n5n}$

Record Ordering. Generally, summaries are reported in a particular order. First, the winning team is reported, then the loosing team, and then the winning team's best player and so on. To better capture such structure, we introduce a new field *"rank"* with the help of which we can order entities according to the way they should be transcribed. To enable this, we use a novel LSTM based pointer network. The LSTM model helps in modeling the sequence information when we pass the input content in a particular order.

The embeddings generated in Sect. 3.3 are fed into a Bi-LSTM encoder and a pointer network decoder which generates ordered indices of the selected records.

Step 2: Summary Generation. For summary generation, we use the encoder-decoder architecture with an attention mechanism. The model takes the record selection output and trains a new model without re-using the embeddings learnt during stage 1. Also we do not want to overload the summary generation model with rank information. The purpose of rank information was just to provide us the optimal set of ordered records for summary generation. We found that reusing the embeddings from the selection stage only added noise to the generation. Our summary generation model is trained to maximize log likelihood of gold output text y given the selected records r from phase-1. Example of summaries are shown in Table 1 and Table 2.

$$max \sum_{r,y \in D} log\, p(y|r) \tag{2}$$

4   Results and Discussion

We conducted our experiments on an Intel Xeon(R) computer with E5-2697 v2 CPU and 64GB memory, running Ubuntu 14.04. We evaluated ROC Bot on

Table 2. Table and Summary from Clinical domain

Employee ID	Employee name	Age	Gender	Salary
1001	Kartik	20	M	10000
1002	Michael	25	M	20000
1003	Peter	30	M	30000
1004	Raj	35	F	40000

With order model Summary: There are total 4 employees. Average age of employee is 27.5. There are 3 males and 1 female. Highest salary is 40000. Lowest salary is 10000. Average salary is 25000.

Without order model Summary: There are total 4 employees. Average age of employee is 27.5. There are 3 males and 1 female. Highest salary is 40000. Lowest salary is 10000. Average salary is 25000.

Table 3. Data sets and their statistics

Data set	# Tables	# Columns	Train. set	Test set
Ent. Data set-1	5	42	1700	300
Ent. Data set-2	3	70	1700	300
WikiSQL	24241	1542564	56355	15878

three real-world data sets, out of which two are our internal enterprise data sets (See Table 3). First data set is related to employees in a large software services company and their allocations. Our second internal data set is about a large pharmaceutical company's stores and employee's information.[1] Our third data set for experimental evaluation is widely used WikiSQL dataset. Ent. Dataset-1 & 2, composed of following type of queries: factoid, self joins, group by, sorting, and aggregations. WikiSQL data set is composed of factoid, aggregation. Queries in WikiSQL were only over single columns.

Note that, in order to generate the intermediate form(sketch), all models are not necessary. The predicate, value, aggregation & operator models are sufficient to generate the sketch. The type, attribute and PV binding models are just used to improve the accuracy. As seen in Table 4, ROC Bot has shown a very good performance for Ent. Data set-1 & 2. Recent works have reported approx. 80% accuracy on WikiSQL data set [3,14]. Our results also demonstrate similar results with simple network architecture. Our entity discovery model was based on the **Lucene** search. It could be further improved by combining **Lucene** search with deep learning based entity extraction. Interestingly, for evaluating Ent. Data Set-2, we just used the same model that was trained on Ent. Data Set-1. We can see that almost similar results were achieved on the latter data set demonstrating the transfer learning capabilities of our ML models.

Table 4. Accuracy across models for all datasets

	Ent. Data Set-1	Ent. Data Set-2	WikiSQL
Predicate Acc.	94.6	93.9	91.4
Value Acc.	94.56	92	88.2
Type Acc.	84	82.86	NA
Attribute Acc.	80	78.2	NA
PV Binding Acc.	92.57	92.34	NA
Aggr. & Operator Acc.	93.1	93.4	92
Overall Sketch Acc.	86.64	84.74	72.1
Overall Exec. Acc.	91.78	87.3	75.24

---

[1] These data sets are available for download and public use at https://github.com/nlpteam19/ROCBot.

5   Conclusion and Future Work

We proposed ROC Bot as a novel framework for performing natural language question answering over BI data. It has capabilities to connect multiple dashboards and provide a single view of the data. The user interface can be used even by naive users who are not familiar with database management. The tangible needs are ease of use in data analytics, insights & reporting. For energy management, such type of capabilities is very important. At the moment the main limitation of ROC Bot is the lack of coverage to explain results to the user and ways to correct the queries if the translation was inaccurate. We plan to add a explanation framework in ROC Bot that would be able to explain the outcomes.

Our approach is based on deep learning using multiple sequence tagging networks and knowledge graph that uses minimal training data and supports data assistance across multiple domains. Our framework captures the user context and provides a robust conversational interface for getting insights in enterprise data. In future, we plan to explore and evaluate multi-tasking capabilities i.e. having an intermediate representation and supporting a range of other tasks.

References

1. Dong, L., Lapata, M.: Coarse-to-fine decoding for neural semantic parsing. CoRR abs/1805.04793 (2018)
2. Ehrenbrink, P., Osman, S., Möller, S.: Google now is for the extraverted, cortana for the introverted: investigating the influence of personality on IPA preference. In: Proceedings of the 29th Australian Conference on Computer-Human Interaction, OZCHI 2017, pp. 257–265. ACM, New York (2017). https://doi.org/10.1145/3152771.3152799
3. Guo, D., et al.: Question generation from SQL queries improves neural semantic parsing. CoRR abs/1808.06304 (2018)
4. Jadeja, M., Varia, N.: Perspectives for evaluating conversational AI. CoRR abs/1709.04734 (2017)
5. Kukich, K.: Design of a knowledge-based report generator. In: 21st Annual Meeting of the Association for Computational Linguistics, pp. 145–150. Association for Computational Linguistics, Cambridge, June 1983. https://doi.org/10.3115/981311.981340, https://www.aclweb.org/anthology/P83-1022
6. Li, F., Jagadish, H.V.: Constructing an interactive natural language interface for relational databases. Proc. VLDB Endow. 8(1), 73–84 (2014). https://doi.org/10.14778/2735461.2735468
7. McKeown, K.R.: Text Generation: Using Discourse Strategies and Focus Constraints to Generate Natural Language Text. Cambridge University Press, Cambridge (1985)
8. Parks, M.: Microsoft Business Intelligence. POWER BI. CreateSpace Independent Publishing Platform, USA (2014)
9. Pasupat, P., Liang, P.: Compositional semantic parsing on semi-structured tables (2015)
10. Peters, M.E., et al.: Deep contextualized word representations. CoRR abs/1802.05365 (2018)

11. Popescu, A.M., Etzioni, O., Kautz, H.: Towards a theory of natural language interfaces to databases. In: Proceedings of the 8th International Conference on Intelligent User Interfaces. IUI 2003, pp. 149–157. ACM, New York (2003). https://doi.org/10.1145/604045.604070, http://doi.acm.org/10.1145/604045.604070
12. Reiter, E., Dale, R.: Building applied natural language generation systems. Nat. Lang. Eng. **3**(1), 57–87 (Mar 1997). https://doi.org/10.1017/S1351324997001502, https://doi.org/10.1017/S1351324997001502
13. Sadun, E., Sande, S.: Talking to Siri: Learning the Language of Apple's Intelligent Assistant, 2nd edn. Que Publishing Company, Indianapolis (2013)
14. Sun, Y., et al.: Semantic parsing with syntax- and table-aware SQL generation. In: ACL (2018)
15. Tian, R., Narayan, S., Sellam, T., Parikh, A.P.: Sticking to the facts: confident decoding for faithful data-to-text generation (2019)
16. Xu, X., Liu, C., Song, D.: SQLNET: generating structured queries from natural language without reinforcement learning (2017)
17. Yaghmazadeh, N., Wang, Y., Dillig, I., Dillig, T.: SQLizer: query synthesis from natural language. Proc. ACM Program. Lang. 1(OOPSLA), 63:1–63:26 (2017). https://doi.org/10.1145/3133887, http://doi.acm.org/10.1145/3133887
18. Yin, P., Lu, Z., Li, H., Kao, B.: Neural enquirer: learning to query tables with natural language (2015)
19. Yu, T., Li, Z., Zhang, Z., Zhang, R., Radev, D.: TypeSQL: knowledge-based type-aware neural text-to-SQL generation. In: Proceedings of the 2018 Conference of the North American Chapter of the Association for Computational Linguistics: Human Language Technologies, vol. 2 (Short Papers), pp. 588–594. Association for Computational Linguistics (2018). http://aclweb.org/anthology/N18-2093
20. Zhong, V., Xiong, C., Socher, R.: SEQ2SQL: generating structured queries from natural language using reinforcement learning (2017)

Digitize-PID: Automatic Digitization of Piping and Instrumentation Diagrams

Shubham Paliwal$^{(\boxtimes)}$, Arushi Jain, Monika Sharma, and Lovekesh Vig

TCS Research, Delhi, India
{shubham.p3,arushi.jain,monika.sharma1,lovekesh.vig}@tcs.com

Abstract. Digitization of scanned Piping and Instrumentation diagrams (P&ID), widely used in manufacturing or mechanical industries such as oil and gas over several decades, has become a critical bottleneck in dynamic inventory management and creation of smart P&IDs that are compatible with the latest CAD tools. Historically, P&ID sheets have been manually generated at the design stage, before being scanned and stored as PDFs. Current digitization initiatives involve manual processing and are consequently very time consuming, labour intensive and error-prone. Thanks to advances in image processing, machine and deep learning techniques there is an emerging body of work on P&ID digitization. However, existing solutions face several challenges owing to the variation in the scale, size and noise in the P&IDs, the sheer complexity and crowdedness within the drawings, domain knowledge required to interpret the drawings and the very minute visual differences among symbols. This motivates our current solution called *Digitize-PID* which comprises of an end-to-end pipeline for detection of core components from P&IDs like pipes, symbols and textual information, followed by their association with each other and eventually, the validation and correction of output data based on inherent domain knowledge. A novel and efficient kernel-based line detection and a two-step method for detection of complex symbols based on a fine-grained deep recognition technique is presented in the paper. In addition, we have created an annotated synthetic dataset, *Dataset-P&ID*, of 500 P&IDs by incorporating different types of noise and complex symbols which is made available for public use (currently there exists no public P&ID dataset). We evaluate our proposed method on this synthetic dataset and a real-world anonymized private dataset of 12 P&ID sheets. Results show that Digitize-PID outperforms the existing state-of-the-art for P&ID digitization.

1 Introduction

A Piping and Instrumentation Diagram (P&ID) is a standardized schematic illustration used in the process engineering industry to record mechanical equipment, piping, instrumentation and control devices employed in the physical implementation of a process. P&IDs are created at the design stage of the process, stored in an image or PDF format and play an important role in the

© Springer Nature Switzerland AG 2021
M. Gupta and G. Ramakrishnan (Eds.): PAKDD 2021 Workshops, LNAI 12705, pp. 168–180, 2021.
https://doi.org/10.1007/978-3-030-75015-2_17

maintenance and modification stage of the physical process flow. Over the years, there are millions of PID sheets that have been manually generated, scanned and stored as images. The valuable information trapped in these images needs to be unlocked and integrated with modern smart P&ID systems. This digitization is necessary to facilitate easy reuse of data and design, automate mundane tasks, maintain inventory, reduce time, increase efficiency and productivity. Currently, P&ID sheets are manually processed by engineers which is a very burdensome, time consuming and error-prone task. There is a very high cognitive load involved in manual digitization due to the minor variations in symbols, scale, size and noise within the sheets, in addition to the crowdedness of text, symbols and line. There is also significant domain knowledge involved in determining line changes and associating text with lines and symbols. Extraction and analysis of textual information, pipelines, and symbols as graphic objects and shapes are the key tasks for interpreting P&ID sheets. We exploit the recent advances in deep learning/machine learning for these tasks.

Several approaches have been proposed for digitizing P&ID sheets or similar documents. This includes conversion of scanned engineering drawings into 3D representation CAD files [10], symbol recognition [3] and classification [1], and shape representation [17]. Ishii et al. [8] presented work towards reading hand drawn piping and instrument diagram where lines, symbols and characters are separated hierarchically from the vectorized representation. In another paper by Gellaboina et al. [7], an iterative learning approach based on hopfield neural networks was presented to detect symbols in P&ID sheets.

Over the last decade, researchers have applied dynamic programming, machine learning, deep learning and pattern recognition to automate the detection of lines, text, shapes from PDFs and/or scanned images. Nazemi et al. [13] presented a method for detecting and extracting mathematical expressions, alphanumeric symbols to generate MathML of the scanned documents. A thorough review of prior methods and a general framework for the digitization of complex engineering diagrams was proposed by Moreno-Garcia et al. [12]. Fu et al. [6] described a visual recognition approach by leveraging CNNs for symbol recognition and methods like multi-scale sliding window and connected component analysis for automatic localization. A semi-automatic and heuristic based approach for symbol localization is proposed by Elyan et al. [4] which utilizes machine learning models like Random Forests, Support Vector Machines (SVM), and CNNs. Kang et al. [9] proposed a two-fold method comprising of extraction of relevant components from P&IDs followed by a recognition step that compares the input sheet at various angles with the objects registered in the database. Very recently, Rahul et al. [14] proposed a novel end-to-end approach based on a combination of low-level vision techniques and deep learning networks like CTPN [15] and FCN [11] for digital interpretation of P&ID sheets by yielding the process flow in a tree format. The shortcoming of approaches proposed in [14] is that it utilizes a hough transform for detecting lines which is parameter-dependent and does not perform well on noisy P&IDs. Moreover, it uses CTPN for text detection which is not able to identify vertical text components present in P&IDs.

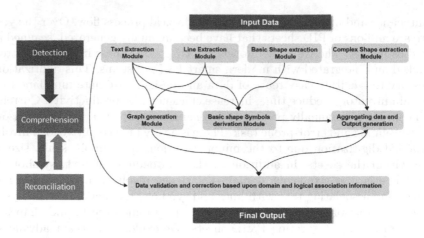

Fig. 1. An overview of Digitize-PID which consists of 3 sequential modules with their corresponding sub-modules: Detection, Comprehension and Reconciliation.

Although significant efforts have been made to improve the performance of automatic methods for conversion of P&IDs into digital drawing, but perfect automatic recognition is still not achievable [12]. To this end, we propose an end-to-end pipeline called *Digitize-PID* which leverages computer-vision techniques and deep learning methods to first detect various components of interest such as lines, graphic symbols and textual information; followed by their aggregation and association with each other; and finally, validation and correction of the extracted output data based on domain rules. We describe a robust kernel-based approach for line detection which works well even in noisy environments. Additionally, a two-step process for detecting complex symbols having minute differences in visual structure is presented which utilizes a deep learning based network for symbol localization and fine-grained classification. We evaluate the effectiveness of our proposed solution on a real-world dataset of 12 P&ID sheets and show impressive results. Note that while 12 may seem like a small number, each P&ID sheet is a very high resolution image with hundreds of visual and textual components. Since, there exists no publicly available dataset for P&ID sheets, we synthesize our own synthetic dataset named *Dataset-P&ID*[1]. We also benchmark this dataset using Digitize-PID and make it publicly available for accelerating community advances in this field. To summarize, our key contributions in this paper are:

- We propose Digitize-PID, an end-to-end novel and robust pipeline for digitizing P&ID sheets by leveraging computer vision and deep learning.
- Digitize-PID combines novel image-processing techniques for hard low-level vision problems such as line detection, dashed line detection, corner detection and a deep learning pipeline for symbol detection and recognition.

---

[1] https://drive.google.com/drive/folders/1Br09_gOKkHsxBOZxH9ojxrnJaxHn333P.

- We create a synthetic dataset of P&ID sheets called Dataset-P&ID consisting of 500 P&ID sheets with corresponding annotations for training and evaluation purposes. The dataset is released online for public use.
- We benchmark our proposed solution Digitize-PID on two datasets: a real-world dataset of 12 P&IDs and a synthetic Dataset of 100 P&IDs, and present the results in Sect. 4.
- We also compared the performance of Digitize-PID against prior state-of-the-art methods by Rahul et al. [14] and outperformed it.

The remaining sections of the paper are structured as follows: Sect. 2 describes the problem statement and discusses about the detection, comprehension, and reconciliation steps of our proposed pipeline. Section 3 provides details about the synthetic dataset that we have generated for training and evaluation purposes. This is followed by experimental details and results in Sect. 4. Finally, we conclude the paper in Sect. 5.

2   Proposed Method: Digitize-PID

In this paper, the task is to automate the process of P&ID digitization to convert the scanned legacy P&ID sheets into a structured format. The proposed method should be capable of identifying different industrial components such as symbols, pipes along with their labeled text and neighbouring symbols. The proposed pipeline *Digitize-PID* takes an input P&ID image and outputs a .csv file consisting of two separate tables listing - (1) different instances of symbols with their mapped text labels and connected pipelines; and (2) containing the list of inter-connectivity between different pipelines representing a graph.

Digitize-PID consists of three high level steps: Detection, Comprehension and Reconciliation, as shown in Fig. 1. The *Detection* step involves extraction of different components from P&ID sheets such as text, lines and symbols which are essential to execute the subsequent steps in the pipeline. The *Comprehension* step consists of logically aggregating the different components detected in the previous stage, for example, the graph generation module takes basic features like lines, symbols and textual information as input and associates the appropriate symbols and text to lines. Finally, the *Reconciliation* step comprises of applying different domain/business rules and final corrections/tweaks on the output data of the comprehension stage. Next, we present a detailed description of these 3 steps in the following sub-sections.

2.1   Detection

The detection module comprises of the following sub-modules, as shown in Fig. 2, which are independent and executed in parallel.

- **Text Extraction module**: P&ID sheets contain text for labeling different components and specifying different parameters of pipelines. We perform text extraction via a 2-step process which involves dividing a P&ID image into

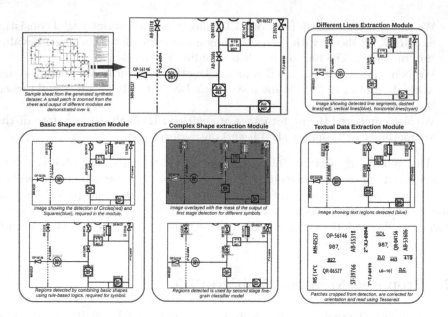

Fig. 2. Figure illustrating different sub-modules of the Detection step of Digitize-PID over a sample P&ID sheet (a small image-patch is zoomed for visual clarity).

multiple fixed-sized overlapping patches. These patches are processed using a Character Region Awareness for Text Detection (CRAFT) [2] network, which predicts bounding boxes (Bbox) for text regions. CRAFT works robustly even for vertically aligned texts. The overlapping Bboxes across overlapping patches are merged using IOU metric which helps to localize the text with high accuracy and effectively reduces the cases of missing texts. These merged Bboxes are projected on the input P&ID sheet and text-patches are extracted. These patches contain single-lined texts read using Tesseract.

- **Line Extraction module**: A P&ID sheet utilizes a network of different types of lines to denote connections between different components, which collectively represent the desired process flow. In Digitize-PID method, we perform line detection using filters based on a structuring element matrix. In the pixel representation, a line can be defined as set of continuous adjacent points in a particular orientation (line orientation). Thus, even an infinitesimal segment of a line can be seen as a basic building block for the entire line. A structuring element is defined as a binary matrix of a fixed dimension $(m \times n)$, in which all active regions denote the filtering line's infinitesimal segment. However, for practical purposes, we do not choose an infinitesimal segment for a structuring element matrix, rather we choose a size greater than the line width and as a function of image spatial resolution, so as to avoid noise and scaling effects in line detection.

Formally, lets assume a binary image $\boldsymbol{A}$ as an integer grid $\boldsymbol{Z}^d$ of dimension d (here $d = 2$), and $\boldsymbol{B}$ is the line structuring element belonging to the same set $\boldsymbol{Z}^d$. We first perform erosion on $\boldsymbol{A}$ using $\boldsymbol{B}$, as given in Eq. 1. As a result, we filter out all the elements not resembling $\boldsymbol{B}$. Next, the filtered regions in the image are restored by performing a dilation operation, as given in Eq. 2, using the same structuring element $\boldsymbol{B}$.

$$\boldsymbol{A}^{erode} = min_{(x',y'):B(x',y') \neq 0} \boldsymbol{A}(x + x', y + y') \tag{1}$$

$$\boldsymbol{A}^{dilate} = max_{(x',y'):B(x',y') \neq 0} \boldsymbol{A}^{erode}(x + x', y + y') \tag{2}$$

$$conv(P) = \left\{ \sum_{i=1}^{n} \lambda_i p_i \mid \sum_{i=1}^{n} \lambda_i = 1 \wedge \forall i \in \{1, .., n\} : \lambda_i \geq 0 \wedge p_i \in P \right\} \tag{3}$$

Subsequently, the pixels obtained in the activated regions generate different contours over line regions. Each contour formed over set P, containing n pixel points, is bounded by a convex hull $conv(P)$. The convex hull $conv(P)$, as defined in Eq. 3, is the intersection of all convex supersets of P [16], which ensures a tight bound over the convex contour of the line. Finally, the two extreme end points from the set $conv(P)$ are computed, along the orientation of the structuring element and are treated as end points of the detected line.

Dashed Line detection: Here, we are focusing on Dashed Lines present in P&IDs which are a series of line segments separated by equal distance, as shown in Fig. 2. We leverage the collinearity and consistency properties of dashed lines for detection. There are two thresholds that we use for segment-length and distance between consecutive line-segments (gaps) which are determined based on the average value for the line cluster having the least mean segment-length and gap. The cases of jumps in the series are very often noticeable in P&ID sheets which lead to inconsistency in gaps between segments of dashed lines. This consistency is retained by applying a rule for filtering out contiguous jumps (three or more). The detected series are then merged to form the continuous chain. The only candidates for merging are the series of segments with opposite orientation and in close proximity with each other which are obtained using the DBSCAN [5] algorithm.

- **Basic Shape extraction module**: Among various symbols used in P&ID sheets, certain symbols are composed of primitive shapes, such as rectangles and circles (Fig. 3). Some of these symbols are differentiated via the texts written inside them. One such basic shape is a circle which are detected by applying Hough transforms across different overlapping image patches followed by their aggregation.
- **Complex Shape extraction module**: P&ID sheets also contain very complex symbols whose structures have very minute inter-class differences and are difficult to interpret and derive via traditional image-processing. These symbols are detected using a 2-step process which consists of a deep learning pipeline for symbol localization followed by fine-grained recognition. As evident in Fig. 3, most of the complex symbols have very similar shape, thus it

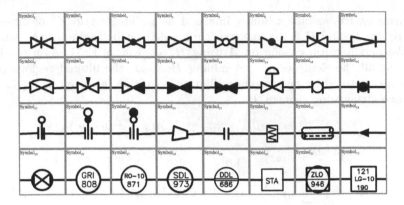

Fig. 3. Figure showing a set of 32 different symbols used for Dataset-P&ID. $Symbol_1$ to $Symbol_{25}$ are complex symbols as they are structurally very similar and are detected using a Complex Shape Extraction module. Remaining $Symbol_{26}$ to $Symbol_{32}$ are detected using the Basic Shape Extraction module.

is preferable to create a common class for all such symbols for symbol localization. For this, we have trained an FCN [11] based semantic segmentation model which is used to localize all such symbols. We apply this FCN model to obtain region-proposals for symbols which are subsequently fed as input to a TBMSL-Net [18] network trained for fine-grained symbol classification.

2.2   Comprehension

Now, we describe how we derive many essential properties of P&IDs by using the appropriate logical combination of text, symbols and lines obtained previously.

– **Graph generation**: The interactions between different components of P&ID are represented by a web of lines that can be interpreted as a weighted graph structure. The graph representation assumes that the components are vertices and the connecting lines between components are edges. The connecting edges are of varying shapes, which can be decomposed into a combination of multiple straight lines of arbitrary lengths which enables us to create a graph with all straight line edges. For graph generation, we utilize the line information extracted in the detection step to filter out the lines with length smaller than the resolution-dependent threshold(α). Similarly, we remove the lines overlapping with regions of texts/symbols. The remaining lines (let's say n) are taken as edges, and the two end-points of each line are taken as vertices. This creates n separate graphs, each having 2 vertices. Thus, the junction centers having k lines will have k vertices, occurring in close proximity, and another k vertices pointing away in different directions. These neighbouring vertices at the junction points can be interpreted as separate clusters such that for each cluster, distance of its respective vertices from its mean (i.e., mean of vertices) would not exceed the line threshold value $\eta\alpha$ (where $0 < \eta < 1$). After optimizing cluster centers with respect to the vertices, we replace the cluster vertices with their respective cluster centers, thus aggregating the separate single-edged graphs to form a common graph.

After the graph structure is created, we assign labels to the edges. Generally, the edges labels are filtered out from the detected text using regular expressions provided by domain experts or manually by visual inspection. After the relevant labels are extracted, they are localized to corresponding graph edges which have the minimum euclidean distance. Finally, these labels on the edges are propagated using Breadth First Search, to the adjoining edges (computed for edges from left to right) with the additional stopping condition of not propagating over label-assigned edges.

- **Basic Shape Symbol detection**: For extracting rectangular shapes, we use the vertex sampling method to obtain candidate vertices of possible rectangular regions, which are later verified as rectangles via their geometric properties. The vertex points are obtained by applying the morphological *AND* operator over the images of vertical and horizontal lines. Further verification of rectangle shapes is done by using the pixel values to satisfy the presence of edges across vertices. Finally, shapes which are different combinations of lines, circles and squares are logically assembled and localized. For cases where multiple symbols have the same shape, we use the embedded-text to differentiate them. These texts circumscribed by the symbols also represent their labels.

- **Data Aggregation**: In this step, different components which include textual, graphical and symbolic information (including both complex and simple symbols) are aggregated such that each detected symbol is mapped to its label, graph nodes, and a separate identification ID is assigned. This helps to create a database of P&ID symbols with their respective properties. Symbol mapping with graph vertices is done by using the nearest neighbour with Euclidean distance as metric. However, a similar approach cannot be used for text boxes as they are of arbitrary length and texts found using the mean will not necessarily be closest. Thus, k-nearest neighbors are computed to get k nearest text boxes corresponding to each symbol. Among these k words, either the regex provided by domain experts are used, or else consistency in the pattern of labels is optimized over all the symbols. The symbol labels pattern which are consistent over other symbol instances in P&ID sheet are finally assigned to the symbol.

2.3   Reconciliation

The digitized data obtained from Comprehension step, are the final output of our proposed method. However, to address any errors/failure, we use the reconciliation step which validates and performs corrections according to domain/business rules. For example, in some arbitrary case, if the particular symbol's label has a static common name over the entire sheet, then the obtained associated text has to be re-validated and overwritten. Multiple iterations involving reconciliation steps can dramatically improve the accuracy of the proposed method even in the customized business scenarios.

3 Dataset

Since, there exists no publicly available dataset for P&ID sheets, we have generated our own synthetic dataset named *Dataset-P&ID* for training and evaluation purposes. Dataset-P&ID consists of 500 annotated P&ID sheets with a 4:1 train-test ratio and is made publicly available for the benefit of research community. It includes 32 different symbols, as given in Fig. 3, uniformly plotted over different graph structures which have been generated to mimic real world P&ID sheets as we have introduced different types of noise such as pixelation, blurring, salt and pepper noise in the generated sheets. The labels are assigned to symbols and pipelines maintaining the standards followed for real world P&ID. The

Table 1. Table showing performance of Symbol Recognition on Dataset-P&ID. **(Bottom-right)** Figure showing the confusion matrix of complex symbols detected using proposed deep learning pipeline on Dataset-P&ID.

(a) Complex Symbols

Symbol	Precision	Recall	F1-score
1	0.932	0.882	0.906
2	0.968	0.968	0.968
3	0.965	0.847	0.902
4	0.974	0.904	0.938
5	0.986	0.973	0.979
6	0.978	0.967	0.972
7	0.971	0.911	0.940
8	0.823	0.963	0.888
9	0.772	0.986	0.866
10	0.974	0.958	0.966
11	0.741	0.991	0.848
12	0.875	0.793	0.832
13	0.972	0.938	0.955
14	0.916	0.961	0.938
15	0.947	0.997	0.971
16	0.979	0.941	0.960
17	0.813	0.979	0.888
18	0.946	0.993	0.969
19	0.946	0.724	0.820
20	0.962	0.929	0.945
21	0.876	0.988	0.929
22	0.936	0.946	0.941
23	0.881	0.956	0.917
24	0.977	0.965	0.971
25	0.927	0.743	0.825

(b) Basic Shape Symbols

Symbol	Precision	Recall	F1-score
26	0.893	0.937	0.914
27	0.864	0.903	0.883
28	0.961	0.975	0.968
29	0.977	0.984	0.980
30	0.890	0.912	0.901
31	0.904	0.892	0.898
32	0.923	0.948	0.935

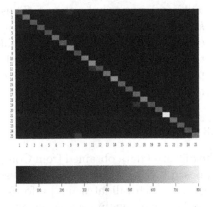

ground-truth of the dataset consists of spatial information of symbols along with associated text labels and their connected pipeline. We provide sets of horizontal and vertical lines with their coordinates and a separate list containing all the texts present in P&ID sheets along with their spatial position.

4 Experimental Results and Discussions

Here, we present the system configuration used for conducting experiments followed by the performance evaluation of Digitize-P&ID. The performance is evaluated based on Recall, Precision and F1-score for different symbols taken over the test-split. A correct prediction of a symbol includes precise localization of symbol with $IOU > 0.75$, symbol class and its associated text-label. Similarly, the output graph is evaluated based on the accuracy of correct adjacency list. However, the validation of graph-creation depends on domain information and is performed as part of the reconciliation step.

(a) **Setup**: We have validated and refined our proposed pipeline via repeated experiments to identify optimal parameters. To begin with, we resize the images to have width of 7168 pixels while maintaining the aspect-ratio. In the detection module, for text detection we split the image into multiple square patches of dimension 800 pixels such that there is an overlap of 50% with their adjacent patches. The common text regions are segregated and read by using Tesseract with line configuration. The same process is also applied on the image by rotating it to capture missing vertical text. Next, we process the entire image at once for line detection. As mentioned earlier, the choice of kernel length is taken as 0.1% of the maximum image resolution. Similarly, in the Basic shape extraction module, the choice of range of radius for hough circle detection is also taken between 0.05% and 0.01% of maximum image resolution. Further, in case of complex shape detection, the image is processed at patch level with patches of size 400px. The output of the FCN model is filtered with threshold probability of 0.8. Finally, the recognition threshold of TBMSL-Net is taken as 0.9 for identifying a symbol from a region-of-interest of image. In the comprehension step, the graph is generated as explained earlier, with the DBSCAN threshold of 50 and the neighbour threshold of 2. The pipeline labels are spread across different pipelines using the Breadth first search algorithm. In data aggregation, we use the standard approach of connecting the nearest line entity. However for texts, we used 5 nearest neighbours, followed by the mapping in accordance with the symbol label rule provided in the reconciliation step. In the absence of such rules, the nearest label texts following the pattern is determined as the associated texts.

(b) **Results & Discussion**: We first present the overall performance of symbol detection with correct associations on synthetic Dataset-P&ID in Table 1a and Table (1b) for the complex and basic shape symbols, respectively. We also show the confusion matrix to demonstrate the robustness of our proposed complex symbol detection module on the synthetic dataset. We use the nearest associated text to resolve the conflict of multiple symbols.

Table 2. Comparison of Digitize-P&ID with prior-art [14] on 12 real-world P&IDs

Symbols	Precision		Recall		F1-score	
	[14]	Ours	[14]	Ours	[14]	Ours
Bl-V	0.925	**0.963**	0.936	**0.986**	0.931	**0.974**
Ck-V	0.941	**0.968**	0.969	**0.988**	0.955	**0.978**
Ch-sl	**1.000**	0.990	0.893	**0.946**	0.944	**0.967**
Cr-V	1.000	1.000	**0.989**	0.973	0.995	**0.986**
Con	**1.000**	0.975	0.905	**0.940**	0.950	**0.957**
F-Con	0.976	0.976	0.837	**0.905**	0.901	**0.939**
Gt-V-nc	0.766	**0.864**	1.000	1.000	0.867	**0.927**
Gb-V	0.888	**0.913**	0.941	**0.946**	0.914	**0.929**
Ins	1.000	1.000	**0.985**	0.964	**0.992**	0.982
GB-V-nc	1.000	1.000	0.929	**0.936**	0.963	**0.967**
Others	0.955	**0.973**	**1.000**	0.990	0.970	**0.981**

Next, we compare our results with Rahul et.al [14]. The symbol detection accuracy is compared on the same set of symbols, used in [14], on the 12 real P&ID sheets dataset as given in Table 2. It shows that the F1-score of complex symbol detection module of Digitize-PID is better than prior art [14]. Please note that the experiment is conducted with all the symbols, but the network is only trained to identify the concerned symbols and the remaining symbols are grouped into an 'others' class. We also illustrate that our proposed structuring element based line detection is more robust than the hough line detection used in [14], as shown in Fig. 4. Finally, we present the performance of text detection and recognition on our dataset in Table 3a. The IOU threshold value is used to find texts having significant overlap with ground truth which are used for further recognition. Since the text-labels contain very critical information, we consider a prediction to be correct only when there is an exact match with the ground-truth label. Table 3b shows the accuracy for line detection, for both complete (99.34%) and dashed (82.91%) lines.

Table 3. Performance of Digitize-PID on synthetic Dataset-P&ID

(a) Performance of Text Detection and Recognition module

IOU	Acc_{Det}	Acc_{Rec}
<0.9	87.18%	79.21%

(b) Performance of Dashed and Complete Line Detection module

Type	Correct	Accuracy
Complete	90774/91416	99.34%
Dashed	20620/24848	82.91%

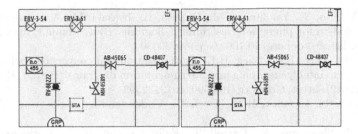

Fig. 4. Left image shows our structuring element based line-detection output and **Right** image shows hough line-detection [14] output.

5 Conclusion

In this paper, we have proposed a complete automated pipeline for extracting relevant information from P&IDs, which are commonly used across several industry verticals for depicting a formal process flow. The proposed pipeline, named Digitize-PID utilizes a combination of state-of-the-art methods for text recognition, robust line detection using morphological operations and a two-step deep-learning based pipeline for fine-grained symbol detection and recognition. Finally, we combine all the extracted information in a graph and organize the extracted data into database-compatible tables. In addition to this, we have synthesized a dataset for P&IDs (Dataset-P&ID) along with their ground-truth annotations which is made public for validation by other researchers.

References

1. Ablameyko, S., Uchida, S.: Recognition of engineering drawing entities: review of approaches. Int. J. Image Graph. **7**, 709–733 (2007)
2. Baek, Y., Lee, B., Han, D., Yun, S., Lee, H.: Character region awareness for text detection (CRAFT). In: Conference on Computer Vision and Pattern Recognition (CVPR) (2019). https://arxiv.org/abs/1904.01941
3. Cordella, L., Vento, M.: Symbol recognition in documents: a collection of techniques? IJDAR **3**, 73–88 (2000). https://doi.org/10.1007/s100320000036
4. Elyan, E., Garcia, C.M., Jayne, C.: Symbols classification in engineering drawings. In: 2018 International Joint Conference on Neural Networks (IJCNN), pp. 1–8 (2018). https://doi.org/10.1109/IJCNN.2018.8489087
5. Ester, M., Kriegel, H.P., Sander, J., Xu, X.: A density-based algorithm for discovering clusters in large spatial databases with noise. In: Proceedings of the Second International Conference on Knowledge Discovery and Data Mining. KDD 1996, pp. 226–231. AAAI Press (1996)
6. Fu, L., Kara, L.: From engineering diagrams to engineering models: visual recognition and applications. Comput. Aided Des. **43**, 278–292 (2011). https://doi.org/10.1016/j.cad.2010.12.011
7. Gellaboina, M., Venkoparao, V.: Graphic symbol recognition using auto associative neural network model, pp. 297–301 (2009). https://doi.org/10.1109/ICAPR.2009.45

8. Ishii, M., Ito, Y., Yamamoto, M., Harada, H., Iwasaki, M.: An automatic recognition system for piping and instrument diagrams. Syst. Comput. Jpn. **20**, 32–46 (2007). https://doi.org/10.1002/scj.4690200304

9. Kang, S.O., Lee, E.B., Baek, H.K.: A digitization and conversion tool for imaged drawings to intelligent piping and instrumentation diagrams (P&ID). Energies **12**, 2593 (2019). https://doi.org/10.3390/en12132593

10. Kanungo, T., Haralick, R.M., Dori, D.: Understanding engineering drawings: a survey

11. Long, J., Shelhamer, E., Darrell, T.: Fully convolutional networks for semantic segmentation. CoRR abs/1411.4038 (2014). http://arxiv.org/abs/1411.4038

12. Moreno-garcía, C., Elyan, E., Jayne, C.: New trends on digitisation of complex engineering drawings. Neural Comput. Appl. **31**(6), 1695–1712 (2019). https://doi.org/10.1007/s00521-018-3583-1

13. Nazemi, A., Murray, I., Mcmeekin, D.: Mathematical information retrieval (MIR) from scanned pdf documents and MathML conversion. IPSJ Trans. Comput. Vis. Appl. **6**, 132–142 (2014). https://doi.org/10.2197/ipsjtcva.6.132

14. Rahul, R., Paliwal, S., Sharma, M., Vig, L.: Automatic information extraction from piping and instrumentation diagrams. In: ICPRAM (2019)

15. Tian, Z., Huang, W., He, T., He, P., Qiao, Y.: Detecting text in natural image with connectionist text proposal network. In: Leibe, B., Matas, J., Sebe, N., Welling, M. (eds.) ECCV 2016. LNCS, vol. 9912, pp. 56–72. Springer, Cham (2016). https://doi.org/10.1007/978-3-319-46484-8_4

16. Weisstein, E.W.: Convex hull. From MathWorld-A Wolfram Web Resource. https://mathworld.wolfram.com/ConvexHull.html

17. Zhang, D., Lu, G.: Review of shape representation and description techniques. Pattern Recogn. **37**, 1–19 (2004). https://doi.org/10.1016/j.patcog.2003.07.008

18. Zhang, F., Zhai, G., Li, M., Liu, Y.: Three-branch and mutil-scale learning for fine-grained image recognition (TBMSL-NET). arXiv preprint arXiv:2003.09150 (2020)

Author Index

Printed in the United States
by Baker & Taylor Publisher Services